LANDSCAPE EVALUATION: APPROACHES AND APPLICATIONS

edited by

PHILIP DEARDEN and BARRY SADLER

Western Geographical Series Volume 25

Department of Geography, University of Victoria
Victoria, British Columbia
Canada

and

Institute of the NorthAmerican West
Victoria, British Columbia
Canada

1989 University of Victoria

LANDSCAPE EVALUATION: APPROACHES AND APPLICATIONS

Western Geographical Series, Volume 25

editorial address

Harold D. Foster, Ph.D.
Department of Geography
University of Victoria
Victoria, British Columbia
Canada

Publication of the Western Geographical Series has been generously supported by the Leon and Thea Koerner Foundation, the Social Science Federation of Canada, the National Centre for Atmospheric Research, the International Geographical Union Congress, the University of Victoria and the Natural Sciences Engineering Research Council of Canada and the Institute of the NorthAmerican West.

Landscape Evaluation: Approaches and Applications
(Western geographical series; ISSN 0315-2022; v. 25)
Includes bibliographical references.
ISBN 0-919838-15-4

1. Landscape. 2. Nature (Aesthetics) 3. Landscape assessment. I. Dearden, Philip. II. Sadler, Barry, 1945- III. University of Victoria (B.C.). Dept. of Geography. IV. Series.
BH301.L3L36 1989 111'.85 C89-091190-8

ACKNOWLEDGEMENTS

This volume reflects the breadth of interest in the field of landscape aesthetics. The editors would like to acknowledge the work of the contributors to the volume, both from the original symposium upon which the volume is based, and also the later solicited papers. Other colleagues also provided helpful advice on individual papers and the overall organization of the volume. Finally we would like to thank the many typists involved with the drafting and redrafting of the articles.

University of Victoria
Victoria, British Columbia

Philip Dearden
Editor

The Western Geographical Series, under my editorship, has now been in existence for twenty years. This milestone could never have been reached without the ongoing support of the University of Victoria, especially that of the Department of Geography. Within the Department, a great deal of hard work and skill is provided by the Technical Services Division, under the direction of Ian Norie. Without the support of the staff employed in this division the twenty-five volumes, which make up the Western Geographical Series to date, would not have been published. In particular I should like to express my appreciation to those who worked on this book. Susan Bannerman undertook the demanding task of typesetting and together with Bruce McDougall was responsible for layout and paste-up. Cartography and photographic production were in the very capable hands of Ole Heggen and Ken Josephson. The efforts of all these individuals are acknowledged with thanks.

University of Victoria
Victoria, British Columbia

Harold D. Foster
Series Editor

PLATE A A ketch in Desolation Sound Marine Park, British Columbia. Scenic beauty is a major source of satisfaction for visitors to the British Columbia coast. ►

TABLE OF CONTENTS

LIST OF TABLES

LIST OF FIGURES

LIST OF PLATES

(All photographs by Philip Dearden)

Our national parks, such as Waterton Lake, are often regarded as the epitome of scenic beauty (**PLATE 1**) and considerable attention is directed toward aesthetics. On the other hand, everyday landscapes are sadly neglected (**PLATE 2**) as in these roadside billboards outside Duncan, British Columbia. ►

Donuts
& Subs
60

EDITORS' INTRODUCTION

PLATE 3 Tarn Hows is one of the best known beauty spots in Britain, yet both lake and forests are totally man-made. ►

1 THEMES AND APPROACHES IN LANDSCAPE EVALUATION RESEARCH

Philip Dearden and Barry Sadler

INTRODUCTION

The visual and aesthetic qualities of the landscape in which we live, work and play make up a fundamental dimension of everyday experience. Our response to them is layered and complex, often difficult to personally capture and articulate, let alone to measure or translate into planning or educational terms. Such a task is the burden of landscape evaluation, and herein lies the source of the fascination and frustration that this field of study holds for those undertaking research and the potential users of their findings. As the subtitle of this volume suggests, the methodological problems of landscape evaluation research constitute the unifying theme of the following papers.

During the past decade, the literature on this subject has grown considerably. Research tends to be trans-disciplinary more than inter-disciplinary, crossing an impressive range of academic and professional boundaries. While geographers have made important contributions to the field, there have been relatively few attempts to take stock of geographical research on landscape aesthetics. The particular dearth of such studies in this country constituted an important supplementary reason for organizing the symposium on which this volume is based. We were curious to find out the range and progress of geographical research being undertaken in Canada.

In this introductory chapter, our purpose is to place the accompanying papers in perspective by reference to the themes and approaches that characterize studies of landscape evaluation. We begin with a summary of some fundamental research issues and questions and their geographical orientations. Next, specific consideration is given to the methodological debate in the field of landscape evaluation. A framework for relating attendant approaches and methods is outlined and then brought to bear on this collection of studies. Finally, landscape evaluation research in Canada is placed in broader geographical and management context by illustrating developments in the United Kingdom and the USA.

A GEOGRAPHICAL PERSPECTIVE ON SOME PERSISTENT QUESTIONS

Landscape evaluation is based on the premise that the aesthetic quality of natural regions, rural countryside and urban areas is a matter of some importance. It incorporates the recognition that the visual characteristics of landscape, like those of buildings and architectural sites, can be systematically analyzed. The nature and scale of this type of inquiry is inherently geographical. By avocation and training, geographers traditionally study the complex and multiform aspects of man-environment relationships. More recently, they have also displayed interest, though not sustained enquiry, in the scenic and aesthetic values which are the transactive currency of physical features and cultural and individual response.[1] It has entailed a reworking, using both humanistic perspectives and behavioral tools, of an earlier, richer seam in geographical thinking, delving into concepts such as sense of place, landscape meaning, and regional identity.[2]

This line of investigation brings geographers into contact with some taxing theoretical and methodological problems. At the core of the field lies a theoretical vacuum that poses still unresolved issues of just what is being examined in studies of landscape evaluation.[3] Such a seeming paradox has long antecedents. It is a result of three centuries of intellectual debate over the nature of the aesthetic. Much more recently the focus of discussion has extended to the relationship of aesthetics and environment, landscape, and place. These should be thought of as a nested set of concepts: landscape is the visible, morphological expression of environment; it connotes a regional assemblage of interrelated physical/cultural features and is thus distinguished by scale from the richer grain and depth of place-bound particularities.

Geographers are well placed to help clarify two major questions underpinning research on landscape evaluation. The first is the issue of aesthetic intentionality in man's role in changing the face of the earth. Landscape making, according to Lowenthal and Prince,[4] is both product and cause of a long succession of visual images and idealized stereotypes. It is important to sort this interplay out, so that the standards for judging the landscape as an aesthetic artifact can be better defined and extended. At present, organizing concepts are coarse grained. We know, for example, that formal gardens in the grand manner are laid out by, and may be judged on, the basis of aesthetic criteria, and that in the poorest and most disadvantaged neighborhoods aesthetics is beside the point. It is the intermediate range of adjustments, where form follows function, often as an unself-conscious vernacular style, that calls for focused analysis.

A second, and related matter, is the man/nature dichotomy in all of this. With the environmental revivalism of the seventies, the aesthetics of land-scape have been fused with the bioethics of nature and ecology. A new *bioaesthetic* has emerged in which the untrammeled play of natural pro-cesses are the focus of aesthetic appreciation. Visual, scenic and related qualities reach their climax in wilderness.[5] Although there is nothing wrong with that *per se*, the corollary is that man-modified and built landscapes tend to be automatically seen in negative terms.[6] While their quality is cer-tainly dubious, the crux of the matter is whether urban and suburban areas are ugly *per se*. This is where geographers and others can turn a deliber-ately aesthetic eye on the man-modified landscape and distinguish the good, the bad and the ugly, indicate the forces at work, and suggest criteria for discrimination.

A start has already been made by authors such as Lowenthal and Prince,[7] Tuan,[8] and Relph.[9] In this volume, Edward Gibson argues con-vincingly that such discussions must be informed by an understanding of the major traditions of landscape aesthetics. These fundamental dimen-sions of considered response and judgement, established by art theory and criticism applied to gardens, architecture, civic design, and landscapes of tourism, comprise a code of seeing. By plumbing persistent tides in taste, Gibson provides a means of setting the spatial geography of landscape aesthetics in more comparative, historical perspective.

The following paper by Dearden lays the groundwork for the methodol-ogical considerations which follow in the second section of the book by examining the factors which underlie landscape preferences. It is a review paper that seeks to identify and organize the factors isolated in the literature as influencing aesthetic landscape assessments. The purpose here is to provide further context, but in this case psychological, as opposed to historical, context. According to Ribe: "As theory formulation becomes more prevalent in aesthetic assessment research, what will be learned of the how and why of people's aesthetic experience cannot help but contribute something to philosophical understanding of landscape aesthetics."[10] We would add "methodological understanding" to his comment: appreciation of the factors influencing judgements of landscape aesthetics is crucial to appropriate research design.

Dearden proposes a pyramid of influences composed of several over-lapping layers with varying degrees of generality amongst populations. At the base of this pyramid are those socio-biological factors common to all human beings. Moving up the pyramid, each succeeding layer — culture, familiarity and socio-economic variables — has more restricted applicability within the population. At the apex of the pyramid, each individual has unique personal characteristics. Thus the scale of enquiry and characteristics of the respondent population have obvious methodological implications, particularly in relation to the degree of consensus on aesthetic values, which forms a major theme in the next section.

ON METHODOLOGICAL CONSIDERATIONS

The evolution of the methodological debate in the field of environmental aesthetics is well documented. On review, one can only wonder whether the authors of early seminal papers, such as Fines,[11] Linton,[12] Litton,[13] Leopold[14] and Shafer *et al.*,[15] largely responding to the need for practical techniques to be applied by practitioners, knew what they were starting. The field, more than any other of which we are aware, has been characterized by ongoing and intensive debate within the academic literature. It is scarcely possible to present even a simple review of the field without eliciting critical accusations of methodological bias.[16] The field is a veritable microcosm of debate regarding the purpose and practice of scholarly enquiry.

There are several reasons why this is so. A major problem relates to the relative newness of the field. Early researchers interested in techniques to assess visual quality, found remarkably little within previous works besides the odd philosophical and historical text, to guide them in this task. Landscape design and art history literature had much to say on aesthetics, but were notoriously difficult to apply to the everyday encompassing landscape that was now the focus of assessment. In short, there were no satisfactory existing theories upon which to develop methodologies.

This theoretical dearth was bemoaned by many, but attacked by few. Instead, concentration focussed upon the development and use of methodologies, often tied only loosely, or not at all, to theories from psychology, geography, art history, socio-biology, landscape design and other areas. In some respects this was understandable. Landscape change was occurring faster than ever before. Practitioners were not going to fiddle with theory while the landscape burned. Many techniques were suggested. They were mostly speculative with little theory nor broad applicability. As a result, initial work in landscape evaluation provided a field-day for the critics.

Not only was the field new, it was also very broad in scope. Researchers and critics came from a wide range of philosophical and methodological backgrounds. It was inevitable that there would be a clash of views regarding philosophies and approaches. Not only were there academics of every stripe and hue but also practitioners — those actually needing to apply the techniques — entering into the debate. There was little commonality between many of those involved, and substantial differences in the object of their deliberations — the landscapes involved and the reasons why the landscapes were being assessed. Every landscape was *sui generis*, each assessor *chaçon à son gôut* and each purpose *ad hoc*. It was a methodological nightmare.[17]

The major philosophical and methodological division has been between those favouring a more reductionist, quantitative-objective approach and those maintaining that it is not possible to apply standard positivist techniques to such a holistic concept as landscape aesthetics. It is not necessary to reproduce all the arguments here, but rather to develop a line of thought that these techniques are not mutually exclusive.[18] It is *not* necessary to decide which one is "best". It *is* necessary to ensure that all techniques conform to the accepted requirements of scholarly enquiry (such as validity and reliability). Given this, all approaches should be able to reveal something of landscape quality. The salient questions are: what do we wish to know, from whom, and for what purpose? Each type of methodology is more or less appropriate for the task depending upon how these questions are answered.

The question of consensus within the population under study has been central to the ongoing methodological debate.[19] Many approaches and methods are based upon the assumption that within society there exists a general consensus on landscape quality; the goal is to assess how this consensus reacts to different landscapes. The focus is to measure landscapes.[20] On the other hand, another body of thought maintains that beauty is in the eye of the beholder; it is, therefore, the eye (or rather the perceptual, affective and cognitive responses) of the observer that should be the focus of attention.[21] Other researchers[22] have proposed a relational view that admits to the presence of both object and subject; landscape appreciation is, therefore, a combination of the two lines of thought.

If this latter view is accepted then any given landscape judgement is a mixture of elements *external* (E) to the observer (i.e. the objects) and *internal* (I) to the observer (the perceptual, affective and cognitive responses). The relative proportions of this mix will vary as a function of the characteristics of the observer(s), the landscape and the mode of interaction. It is not the purpose to discuss the characteristics in depth here, although they are indicated in Figure 1,1 for interest. What is important is the hypothesis relating this mix of internal and external factors to consensus. Specifically, "the potential for societal consensus on landscape quality is directly proportional to the ratio of external to internal influences on the observer. Thus, if E is greater than I, consensus will be high; if I is greater than E, consensus will be low."[23]

The I:E ratio is not an easy one to assess directly. Nonetheless, the literature provides evidence that some circumstances are likely to lead to

CIRCUMSTANCES	EVALUATION COMPONENTS	CIRCUMSTANCES
Surrogates Time Short Appraisal	INTERACTION	Field Time Long Preferential
Young Unfamiliar Trained	OBSERVER	Old Familiar Untrained
Small Area Uniform Extreme Water Natural	LANDSCAPE	Large Area Complex Common No Water Built
High	CONSENSUS	Low
←		→
E>I Objectivist	Evaluation Philosophy	I>E Subjectivist

← **MODEL SUITABILITY** →

ECOLOGICAL	FORMAL AESTHETIC	PSYCHOLOGICAL	PSYCHOLOGICAL	PHENOMENOLOGICAL
eg Leopold [25]	eg Litton [26]	eg Dearden [27]	eg Kaplan [28]	eg Gold & Burgess [29]

FIGURE 1,1 Theoretical Framework Based on Consensus for Landscape Evaluation

E greater than I and others to I greater than E (Figure 1,1) and hence predispose to higher or lower levels of consensus. The significance of this lies in the varying dependencies of different techniques on assumptions of consensus. Hence, some techniques, firmly rooted in an objectivist philosophy, are purely landscape oriented and merely assume consensus. Such techniques might be more suited to the circumstances outlined on the left side of Figure 1,1. On the other hand, other techniques pay little attention to landscape, assume that each observer is unique, there is no consensus and focus their efforts on a subjective analysis of the individual. Where consensus *is* weak (right hand side of Figure 1,1), these techniques might be most appropriate.

In sum, the approaches that have evolved to assess landscape aesthetics should not be seen as mutually exclusive, as they so often have been in the literature. Rather they are complementary; each are more or less appropriate given the purpose and circumstance of the evaluation. Most books in the field have gained their unity from investigating and illustrating just one type of approach. The focus of this collection is the attempt to try to illustrate the broad range of approaches in use and how they all succeed in telling something of the aesthetic quality of the landscape.

A FRAME OF REFERENCE

Daniel and Vining[24] suggest a methodological classification ranging from ecological through formal aesthetic to psychophysical, psychological and phenomenological approaches. Figure 1,1 shows how this classification might be related to the concepts of consensus described above. At one pole, phenomenological approaches, based upon a subjectivist philosophy, seek insight into landscape through in-depth investigation of individual interpretations. These might be based on personal introspection or the interpretations of literary or artistic portrayal of landscape.

The paper by Porteous in this volume illustrates such an approach. It is a subjective interpretation (the author's) of another subjective interpretation (Lowry's) of the landscape of British Columbia. Methodological focus is upon the subject (Lowry), as opposed to the object (landscape), qualitative judgements and skilled interpretation. It is rich in detail and insight conveying to the reader an intimate feel for the landscape in question. It is also highly subjective, probably of restricted validity and reliability and extremely difficult to apply to practical purpose.

Techniques at the other end of the methodological spectrum (Figure 1,1) were designed more specifically with the latter in mind. They are objectivist in philosophy and concentrate upon the landscape as opposed to the observer. Early techniques introduced by Leopold[30] and Linton[31] best exemplify this pole. Although their reliability is likely to be higher than more subjective-oriented techniques, due to the common constraints enforced by the techniques, their validity is more open to question. The techniques might consistently distinguish between landscapes, but is the distinction based upon aesthetic qualities or merely landscape character? That a technique may reliably distinguish the latter does not qualify it to pronounce upon the former. Landscapes may be mapped according to such distinctions (e.g. relative topography, land use) but these have to be linked to aesthetic criteria (such as public preferences) by strong empirical evidence rather than taken as axiomatic.

Recognition of the problem has led to reduced use of this kind of approach although there are circumstances where it might still be used to advantage. Penning-Rowsell, in this volume describes large scale, objectivist, landscape-based research currently being funded by government agencies in the U.K. Here, though, the purpose is to provide data to monitor landscape change, rather than a purely aesthetic concern. Landscape-based techniques might also be appropriate where high consensus levels might be anticipated (Figure 1,1). In such cases the main task is to distinguish between landscapes.

The paper by Moss and Nickling in this volume has elements of this ecological approach in that they concentrate largely upon distinguishing between and mapping landscape features of known scenic quality as opposed to assessing human response. They do, however, identify one of the accepted deficiencies of the ecological approach, namely that it is a static rather than dynamic record of landscape. The authors call for a process-oriented approach and provide several examples of Canadian work in the field to illustrate their concern.

Between these two methodological poles, with concerns either exclusively landscape or subject oriented, lies a range of approaches that try to combine the two elements to a greater or lesser degree. The psychological approaches reflect a stronger subject orientation; the formal aesthetic approaches represent a stronger landscape orientation. The psychophysical approaches straddle the two and attempt to measure both landscape and response in a reasonably comprehensive manner. This probably represents the most favoured methodological approach now being pursued. Several reviews[32] have found psychophysical approaches to be most desirable in terms of their reliability and validity. They are perhaps the most favoured "scientific" approaches (with phenomenological the most favoured "scholarly" and formal aesthetic the most favoured "practitioners" approaches). Psychophysical approaches often make use of large public samples and statistical techniques. They are the most complex and this tends to limit their utility. Many are too complex to be readily understood by practitioners.

In this volume, the paper by Pomeroy, Fitzgibbon and Green represents this line of enquiry. It is based upon personal construct theory, elicited by the repertory grid technique to conduct both similarity and preference analyses. It employs a relatively large number of respondents, relies on complex statistical techniques (particularly multi-dimensional scaling [MDS]), and illustrates the common practice of using colour photographs in this kind of research. The authors demonstrate the techniques by reference to a perceived similarity analysis of the riverscape of the South Saskatchewan River and a preference analysis for the townscape of Wingham in south Ontario. Although methodological details differ between the two studies, the authors conclude that, "the adaptability of personal construct theory, the repertory grid and MDS to the two methodologies while maintaining theoretical validity demonstrates a strength of this approach."[33] Nonetheless, from a utilitarian point of view the use of a "non-metric alternating least squares multi-dimensional scaling"[34] analysis is probably sufficient to discourage many practitioners.

The paper by Marsh also has psychophysical characteristics. Although very different from the preceeding paper, Marsh also attempts to find human response (in the form of photographs taken for postcards) to landscape characteristics (the content of the postcards) in Glacier National Park. A major methodological problem in the field of landscape aesthetics has always been in deriving ways to measure human response to landscape. Many techniques either ignore the problem (e.g. the ecological techniques), obtain in-depth response from few individuals (e.g. the phenomenological techniques) or a more restricted response from individuals (e.g. the psychophysical techniques). Many of the latter tend to be quite obtrusive. Marsh's exploratory paper does not have this disadvantage. It is completely unobtrusive and yet gains response from a large sample.

A number of methodological questions however, are noted by Marsh:

a) how representative is the sample of postcards?
b) were all postcards consumed, i.e. bought?
c) might some postcards have been unpopular due to their content?
d) who bought postcards?
e) why did people buy particular postcards?
f) do the postcards say more about the landscape preferences of the photographers than the public?
g) how can the quantitative data available from the postcards be analysed?

There is no doubt that the postcards can tell us something about landscape, but whether they can be reasonably expected to be indicative of tourist landscape preferences is a difficult question. Marsh finishes his paper with some suggestions for future research in this area.

The formal aesthetic approaches are landscape based but usually attempt to include some elements of subjective response, if only of the person undertaking the survey. To this end they are also known as "expert" techniques due to the reliance upon the judgement and expertise of the person undertaking the survey. They are the most commonly used in practice in North America but have been roundly criticized in terms of both reliability and validity.[35] The paper by Itami in this volume describes much of the work undertaken within this paradigm and the approaches used by three major American agencies, the U.S. Forest Service, the Bureau of Land Management and the Federal Highway Administration. Each has its strengths and weaknesses and Itami describes several comparative studies that have been undertaken that illustrate some of the methodological problems associated with this kind of approach.

The remaining methodological grouping shown in Figure 1,1 comprises psychological or cognitive techniques. This line of enquiry is, as the name suggests, concentrated upon understanding why respondents have preferences for different landscapes. It is, therefore, considerably more theoretically inclined than the other paradigms. This theoretical strength however, is perhaps counterbalanced by the difficulties in applying the findings to actual planning problems. Itami provides a useful short description of some of the work in this area but, more importantly, cites several examples where this line of approach has been combined with other paradigms, particularly the psychophysical and cognitive, to produce results combining the strengths of all three approaches.

The final two papers in the methodolgical section of the volume are much broader in scope. Wood looks at the question of colour in landscape. His exploratory paper spans the methodological continuum discussing the purely subjective responses to colour used by artists through to purely landscape-based colour characteristics. Rather than illustrate a particular methodological line of approach this paper suggests a topic of generic interest, colour, that has been relatively little explored in any of the methodological paradigms discussed earlier. Wood suggests several interesting lines of approach in further developing our understanding of the role of colour in landscape aesthetics.

Hamill also provides a different perspective on landscape assessment methodologies, one that is not paradigm-specific. Hamill suggests that considerable and continuing methodological error exists within the field, even after such error has been demonstrated to exist in the scholarly literature. He discusses several of these perpetuating errors and cites examples of their persistence. His commentary, though directed specifically at techniques for landscape assessment is also applicable to other areas of scholarly work. That it is probably more visible and persistent in the field might be attributable to some of the factors mentioned earlier in this chapter such as the newness of the field, lack of established theory, prevalence of philosophical and methodological debate and the practical imperative to produce results for planning purposes.

A BROADER PERSPECTIVE ON CANADIAN RESEARCH

The distinctive character of the Canadian landscape is drawn by implication from example, rather than explicitly delineated, in this volume. It forms a context for, rather than a focus of, analysis. Although this is understandable given the orientation of the volume, it is work emphasizing by way of conclusion that visual character and aesthetic qualities of the Canadian landscape and the way these are changing, deserve sustained attention for two main reasons.

First of all such a task, we suggest, is a useful focus in its own right as an intellectual endeavour. There is much we do not know about the landscapes of Canada, our responses to them, and what this tells us about ourselves. What qualities and images of space and landscape, for example, are the Canadian equivalents of the features traced by Lowenthal in "The American Scene"?[36] How do Canadian attitudes compare to those in the U.S.A., a country with which we share a common continent and a similar cultural heritage and currency of landscape appreciation? The axial transformation of the Canadian view of landscape, most notably perhaps wrought by the Group of Seven and since extended by the literati, are reasonably well-known but still remain largely unanalyzed as geographies of the mind.

Analyses of the Canadian landscape, secondly, are important, for gaining an understanding of contemporary change. The "New Conservation" of the mid-sixties was stimulated by twin concerns over the reduction in wilderness and open space and the deterioration of the quality of natural and built environments. Aesthetics rather than ecology was the initial yardstick for action. In the United States, this root has continued to grow into a fairly healthy branch of landscape evaluation and visual management as Itami's paper indicates. A related offshoot is evident in the U.K., where the concept of amenity underlies and gives impetus to visual analysis and actions directed at the care of the countryside. Penning-Rowsell's review of landscape research in Britain indicates the volume is small in relation to the task at hand. This point can be underlined with respect to the Canadian situation. The focus and scale of landscape evaluation in this country seems insufficient to provide even a rudimentary basis for visual management in the sense of contemporary practice in the United States.

How can this situation be transformed? First, by geographers and others placing more emphasis on their traditional concern with landscape and the way it is changing in Canada. At a basic level, geographical systems of classification are available to organize discussion on landscape problems. For research purposes, traditional geographical tools and techniques for analyzing the morphology of urban and cultural landscapes are now supplemented by a more eclectic range of approaches, as the papers outlined in this volume illustrate. Secondly, the discipline must do a better job of public and specialized education on these matters. Unless people are alerted to what is happening, especially to the insidious, incremental adjustments through which much visual change occurs, then the basis for action is unlikely. A change in our view of landscape ultimately changes the landscape itself. Third and finally, comes the thorny task of translating analysis into action. Most problems, as Kevin Lynch notes in his seminal book on *Managing the Sense of a Region*[37] are presented poorly, hidden assumptions are often made, implied values are dubious, and the proposed solution is often impossible or at least unrelated to the institutional realities of decision making. This linkage is not one that academic geographers have done a good job of forging, despite the fact that many studies of landscape evaluation (like resource management) are ostensibly undertaken with a policy rationale. We hope that this volume will stimulate thinking on these and related issues, and, above all, give geographers and others a sense of *emerging methodological* challenges and opportunities in landscape evaluation research. Once these are met, we can perhaps move more confidently to the larger tasks involved in the application of findings in planning and design and in the education of those involved in making and remaking the landscape.

REFERENCES

1. For example see the review by DEARDEN, P., "Landscape Assessment: The Last Decade", *Canadian Geographer*, 24, 1980a, pp. 316-325.

2. For example see: LOWENTHAL, D. and PRINCE, H., "The English Landscape," *Geographical Review*, 54, 1964, pp. 309-346; TUAN, Y.F., *Topophilia*. Englewood Cliffs: Prentice Hall, 1974; and RELPH, E., *Place and Placelessness*. London: Pion Press, 1976.

3. APPLETON, J., "Landscape evaluation: the theoretical vacuum", *I.B.G. Trans.*, 66, 1975, pp. 120-123; and DEARDEN, P., "Consensus and a theoretical framework for landscape evaluation", *Journal of Environmental Management*, 34, 1987, pp. 267- 278.

4. LOWENTHAL and PRINCE, 1964, *op. cit.*

5. Although some would dispute this, for example, see WALTER, A., "You'll love the Rockies", *Landscape*, 27, 2, 1983, pp. 43-47.

6. Most studies have found distinct preferences for natural over built landscapes. See, for example, ZUBE, E.H., PITT, D.G., and ANDERSON, T.W., "Perception and prediction of scenic resource values of the Northeast", in ZUBE, E.H., *et al.*, (eds.), *Landscape Assessment: Values, Perceptions and Resources*. Stroudsburg, Pa.: Dowden, Hutchinson and Ross, 1975, pp. 151-168.

7. LOWENTHAL and PRINCE, 1964, *op. cit.*

8. TUAN, 1974, *op. cit.*

9. RELPH, 1976, *op. cit.*

10. RIBE, R.G., "On the Possibility of Quantifying Scenic Beauty — A Response", *Landscape Planning*, 9, 1982, p. 72.

11. FINES, K.D., "Landscape Evaluation: A Research Project in East Sussex", *Regional Studies*, 2, 1968, pp. 41-55.

12. LINTON, D.L., "The Assessment of Scenery as a Natural Resource", *Scottish Geographical Magazine*, 84, 1968, pp. 218-238.

13. LITTON, R.B., *Forest Landscape Description and Inventories*. Berkeley, Calf.: U.S.D.A. Forest Service Resource Paper, PSW-49, Pacific S.W. Forest and Range Experimental Station, 1968.

14. LEOPOLD, L.B., "Quantitative comparison of some aesthetic factors among rivers", *Geological Survey Circular*, 620, 1969, Washington, D.C.

15. SHAFER, E.L., HAMILTON, J.F., and SCHMIDT, E.A., "Natural Landscape Preferences: A Predictive Model", *Journal of Leisure Research*, 1, 1969, pp. 1-19.

16. For example see review by DEARDEN, 1980, *op. cit.*; and replies by HAMILL, L., "Critique of 'Landscape Assessment: the Last Decade' by P. Dearden", *Canadian Geographer*, 26, 1982, pp. 70-72; and HAMILTON, W.G., "Humanistic qualifications of 'Landscape Assessment: the Last Decade' by P. Dearden", *Canadian Geographer*, 26, 1982, pp. 73-76.

17. Early cut and thrust in the field is epitomized by the paper by FINES, 1968, *op. cit.*, and the methodological reply by BRANCHER, D.M., "Critique of K.D. Fines' 'Landscape Evaluation: a research project in East Sussex'", *Regional Studies*, 3, 1969, pp. 91-92. Since this early tussle many other critiques and debates within the literature have taken place. Some examples include: HAMILL, L., "Analysis of Leopold's Quantitative Comparisons of Landscape Esthetics", *Journal of Leisure Research*, 7, 1975, pp. 16-28.; JACQUES, D.L., "Landscape Appraisal: The case for a Subjective Theory", *Journal of Environmental Management*, 10, 1980, pp. 107-113; CLAMP, P., "The Landscape Evaluation Controversy" *Landscape Research*, 6, 1981, pp. 13-15; DEARDEN, P., "Landscape Evaluation: The Case for a Multidimensional Approach", *Journal of Environmental Management*, 13, 1981a, pp. 95-105; POWELL, M., "The Landscape Evaluation Controversy", *Landscape Research*, 6, 1981, pp. 16-18; SHUTTLEWORTH, S., "Landscape Appraisal: the 'Objective/Subjective' debate", *Landscape Research*, 6, 1981, p. 33; RIBE, R., *op. cit.*; CARLSON, A.A., "On the Possibility of Quantifying

Scenic Beauty — Response to Ribe", *Landscape Planning*, 11, 1984, pp. 49-65; and RIBE, R., "On the Possibility of Strong versus Weak Quantification of Scenic Beauty — A Further Response to Carlson", *Landscape Planning*, 12, 1986, pp. 421-429. ITAMI, R.M., "Scenic Perception: Research and Application in U.S. Visual Management Systems", pages 211-241 in this volume provides a good review of critiques undertaken on methodologies used by major American agencies.

18. See DEARDEN, 1981a, *op. cit.*, and 1987, *op. cit.* for amplification of this view.

19. See DEARDEN, P., "Consensus and the landscape quality continuum: a research note", *Landscape Research*, 26, 1981a, p. 31.

20. For example see early papers by LEOPOLD, 1969, *op. cit.*, and LINTON, 1968, *op. cit.*, and the techniques that have derived from these lines of thought.

21. See paper by JACQUES, 1980, *op. cit.*, for an articulation of this point of view.

22. For example see ZUBE, E.H., SELL, J.R., and TAYLOR, J.G., "Landscape Perception: Research, Application and Theory", *Landscape Planning*, 9, 1982, pp. 1-33.

23. DEARDEN, 1987, *op. cit.*, p. 269.

24. DANIEL, T.C., and VINING, J., "Methodological Issues in the Assessment of Landscape Quality", in ALTMAN, I., and WOHLWILL, J.F. (eds.), *Behavior and the Natural Environment.* New York N.Y.: Plenam Press, 1983, pp. 39-84.

25. LEOPOLD, 1969, *op. cit.*

26. LITTON, 1968, *op. cit.*

27. DEARDEN, P., "A Statisitical Technique for the Evaluation of the Visual Quality of the Landscape for Land-use Planning Purposes" *Journal of Environmental Management*, 10, 1980, pp. 51-68.

28. KAPLAN, R., "Some Methods and Strategies in the Prediction of Preferences", in ZUBE, E.H., BRUSH, R.O., and FABOS, J.G.,

(eds.), *Landscape Assessment: Values, Perceptions and Resources*. Stroudsburg, Pa.: Dowden, Hutchinson and Ross, 1975, pp. 118-135.

29. GOLD, J.R., and BURGESS, J., *Valued Environments*. London: George Allen and Unwin, 1982.

30. LEOPOLD, 1969, *op. cit.*

31. LINTON, 1968, *op. cit.*

32. For example see DANIEL and VINING, 1983, *op. cit.*

33. See p. 154 this volume.

34. See p. 158 this volume.

35. For example see KOPKA, S., and ROSS, M., "A Study of the Reliability of the Bureau of Land Management Visual Resource Assessment Scheme", *Landscape Planning*, 11, 1984, pp. 161-166; and MILLER, P.A., "A Comparative study of the BLM Scenic Quality Rating Procedure and Landscape Preference Dimensions", *Landscape Journal*, 3, 1984, pp. 123-135.

36. LOWENTHAL, D., "The American Scene", *Geographical Review*, 58, 1968, pp. 61-88.

37. LYNCH, K., *Managing the Sense of a Region*. Cambridge, Mass.: M.I.T. Press, 1976.

PLATE 4 The side of Meares Island viewed from Radar Hill in Pacific Rim National Park, British Columbia was destined for clearcutting of its forest cover, before protesters stopped logging activities. The case is now before the courts. ►

PLATE 5 The Tantulus Mountains, viewed from Garibaldi Park, British Columbia, epitomize sublime landscapes. ►

PART I

BACKGROUND

PLATE 6 This thatched roof cottage in England would constitute a typical component of a picturesque landscape. ►

2 TRADITIONS OF LANDSCAPE AESTHETICS: 1700 - 1985[1]

Edward Gibson

INTRODUCTION

In 1935, the year that Walter Gropius published his manifesto for modern landscape design, *The New Architecture and the Bauhaus*,[2] Vaughan Cornish published his positivist interpretation of landscape aesthetics, *Scenery and the Sense of Sight*.[3] He thus launched geographers on an intellectual trajectory pointing to a scientific understanding of an ancient geographic question: Why do some places appeal to us more than others? In 1931 Cornish had written a less scientific account of this same question, *The Poetic Impression of Natural Scenery*[4], and eight years later he was to publish his final work on the subject, a pointedly empirical inventory, *The Beauties of Scenery: A Geographical Survey*.[5] His works in landscape aesthetics, although epistemologically distinct, are joined in his total contribution to geography and aesthetics. One can say of geography, since Vaughan Cornish, that it pursues concurrently all three of the approaches to landscape aesthetics utilized by him — humanistic, scientific, and utilitarian.

Well-known humanistic essays by David Lowenthal and Hugh Prince "The English Landscape" and "English Landscape Tastes" appeared in 1964[6] and 1965[7] respectively, at the very moment when so much wilderness and pastoral scenery was being transformed by post World War II industrial and urban expansion — most of it following the 1935 principles of rational design worked out by Walter Gropius. These publications by Lowenthal and Prince had a beneficial effect on the course geographers have taken toward an understanding of landscape aesthetics. They described the visual qualities of scenery affected by geology, vegetation and topography as did Cornish in his survey; but, they gave the man-made landscape far more attention, describing "historic" villages, cities, fields, factories and parks. Lowenthal and Prince noted the spatial arrangement of these elements in the context of atmospheric characteristics and with regard to the "romantic" and "picturesque" attitudes they created. They were notably sensitive to the phenomenology of landscape taste — the variations associated with social class; that is, their own values of conservation and historicism. But from the perspective of art history in 1985 one can understandably identify failures. Lowenthal and Prince failed to define the critical differences between picturesque and sublimity, between the romanticism of eighteenth century landscape arts and the romanticism of nineteenth century landscape arts; and, they gave no attention to the new British "pop" designs and tastes that were being formulated and implemented around them.

In his 1975 book, *The Experience of Landscape*[8], Jay Appleton gave geography a work that was both more encompassing and systematic in its treatment of art theory and at the same time more restrictive in its reductionist dependence on socio-biology. Appleton's geographic background did not preclude him from bringing a systematic review of art theory — symbolism, painting, architecture, literature, civic design and photography — to bear on the problem of aesthetic potential of places. But Appleton, like Lowenthal and Prince before him, failed to examine the social production of reality, how society reproduces itself, and the cultural production of landscape aesthetics — the practice of art, including the commerical arts of film and television.

Edward Relph's study of *Rational Landscapes and Humanistic Geography*[9] caught the historical drift of contemporary debate in landscape design; his criticism of modernity paralleled the crisis in rational landscape design presented three years before in the first edition of *The Landscape of Post-Modern Architecture* by Charles Jencks.[10] However, he did not plumb the depths of the debate between humanism and rationalist aesthetics. Relph omitted a critique of modernity. He chose instead to dismiss it, positing in its place an aesthetic of landscape that can only be said to be reactionary, in precisely the same sense that Daniel Bell's thesis on the post-industrial city is reactionary.

If the legacy of Cornish's humanistic work on landscape aesthetics has been carried forward by Relph's grasp of historical moment, other Canadian geographers have furthered a scientific understanding of the subject by keeping an ear open to local politics and an eye open to local turf. The growing interest in a science of landscape aesthetics begun half a century by Cornish is continued in the works of Philip Dearden,[11] Louis Hamill,[12] William Hamilton,[13] Steven Pearce[14] and Nigel Waters.[15] A major reason for these geographers being at the advance edge lies in the fact that their turf, Alberta and British Columbia, comprises some of the most talked about scenery in the world. This landscape is being encroached upon by public and private resource projects in the Rocky Mountains, on the Pacific Coast, and along the eastern foothills. Corporations want to transform crown lands into ski resorts; utility commissions want to stretch power and pipe lines straight across mountain crests; cottagers and naturalists want to preserve vistas; city hikers want to retreat to wilderness; people with motor-homes want parts of the city conveniently dispersed in the forest; and coal tycoons and lumber barons want to dig up mountains and clear-cut panoramas. Few of these interests are compatible with others.

The legal and administrative requirements of environmental impact studies combined with public hearings reflect demands for ways of evaluating potential landscape changes and allocating standards for scenery in some democratic way. The stakes are high and so are the levels of emotional commitment — hence the need for an objective way of uncovering the variables that underlie public preferences for scenery. All to the good; but an additional body of evidence of what people like about scenery — natural and man-made — is found in the history and theory of landscape arts. This comprises landscape painting, gardening, architecture, civic design, and sight-seeing and it includes contemporary art theory and criticism.

The argument of this chapter is that all geographers working in landscape aesthetics, although not limited by the conventions of art history and theory, are indebted to both. This debt is at present attributed to only a narrow selection of possible historical periods and theories — mainly the picturesque theory of eighteenth century England and the rationalist theory of contemporary consumer capitalism. The proper concern for geographers who seek to understand why some scenery is more appealing than others is the rich legacy of aesthetic traditions that has evolved over the past four centuries. These traditions are distinct. They tend to replace one another as the dominant tradition through time and between human groups. But they also overlap and exhibit a broad historical consistency.

There are five aesthetic traditions whose identification by geographers promises to enrich both humanistic scholarship and scientific research on landscape aesthetics. In order of their review these traditions are: 1) the rational tradition; 2) the picturesque tradition; 3) the sublime tradition; 4) the realist tradition; and 5) the surrealist tradition.[16]

RATIONAL TRADITION

Central to Western aesthetics are the notions of rational, spatial arrangement of visual unity, hierarchy and symmetry, and of mental associations with the utility of places and objects visualized. The appeal of rationalism began in the Italian renaissance during the fifteenth century and diffused to

Spain, France, the Low Countries, the United Kingdom and then throughout Western Europe and parts of the New World. The appeal was universal until the end of the eighteenth century. It was felt in civic design, in the layout of natural parks, gardens, architecture and in landscape painting. It produced systematic, symmetrical arrangements of vegetation; it produced arrangements of perfect forms, squares, circles, rectangles, and oval shapes, axial avenues stretching to the horizon and infinity; and it led to the structuring of vistas in foreground, middle ground and background. Yet, in the passion for geometric scenery there were constraints.

In his popular guide to landscape design, *Vitruvius Britannicus*, 1717,[17] Colin Campbell followed Alberti and instructed British designers to scale buildings and gardens so that they corresponded to, or harmonized with, the surrounding scale of topography. Elsewhere, people were instructed to scale their geometric patterns to correspond to human vision; city blocks and garden components in park layouts were to be limited to distances at which the eye could recognize the identity of a human face. There were, of course, variations on the rationalist theme and at times, in the Baroque phases of seventeenth century Britain and eighteenth century France, for instance, human scale was eclipsed by themes of constriction and release and by the "super" human scales of avenues, squares and vistas.

The theoretical basis for this rational aesthetic was Plato's theory of ideas and scholastic realism. In seeing "perfect" forms, ideas of abstract perfection (pure mathematics) and hence God were to spring to mind — therein lay the source of this aesthetic feeling.

While the universal appeal of rationalism gave way in literature, painting, gardening and architecture to the theme of the picturesque during the eighteenth century, by the beginning of the twentieth century a modern theme, a variant of the rational tradition, began to prevail.[18] This is perhaps best encapsuled by Walter Gropius in *The New Architecture and the Bauhaus*. Like the renaissance theories before them, modern theories were based on realism, but scientific realism not scholastic. They were based on the notion of universal (mathematical and statistical) truth, on proportion of parts, and they produced an appeal for perfect forms, squares, rectangles and circles.

The modern equivalent of rational aesthetics had elements that contrasted with earlier rationalism. Calculated asymmetry, as distinct from symmetry and randomness, is a guiding principle in massing buildings, arranging natural vistas in civic design, gardening, photography, and abstract painting. The effects of scientific realism — new dimensions of social production associated with Albert Einstein's theory of relativity, with X-ray technology, and with Neil Bohr's theory of atomic order — have found their way into cultural production. They entered art theory through cubist painting, then spread to architecture, gardening and civic design. The effects in landscape arts are manifest in the appeal of picture-plane two-dimensional, not three-dimensional, visual perspective, flat surfaces, overlapping planes, glass see-through walls and vegetation screens. They are manifest in buildings and masses of vegetation that appear to float in space, an effect produced by building on uncovered posts and placing water in the foreground.

Unlike the renaissance forms of rationalism, modern rationalism has not intended a metaphysical basis. Bauhaus designs *did* intend to liberate society from the tyranny of work and the dehumanizing influence of industrial capitalism; but the buildings they produced *were not* intended to say that they were *not about* anything in the sense that renaissance styles were *about* the Holy Kingdom, heaven-on-earth. The antagonism Edward Relph[19] holds toward the results of modern rational design theory seems based on the meaning of international capitalism, international communism, and naive scientism which we now, and after the fact, associate with the Bauhaus and International styles.

The tradition of rationalism in landscape aesthetics is an abiding theme in Western thought. It prevailed for several centuries after 1400 then faded from popularity only to re-emerge as a dialectic of other competing traditions. Notable among the aesthetic traditions through which the modern variation of rationalism has been transformed is the tradition of the picturesque.

PICTURESQUE TRADITION

The picturesque tradition, like the rational tradition, is deeply entrenched in Western landscape arts, particularly after the seventeenth century. Like all aesthetic traditions, its rise to dominance and its persistence are understood by placing its popular appeal in the context of the ways society reproduces itself. The theory of the picturesque emerged in the eighteenth century in the context of the fact or myth of the autonomous individual that developed with British liberalism, the theory of capitalism, and the moral convictions of civic humanism that followed the Glorious Revolution in England. The character of the picturesque theory and its results in architecture, civic design, literature and especially garden design were influenced by the literate bourgeoisies, by landscape painting, and, by sight-seeing in Italy. These combined to shape the social reality of English-speaking peoples around the world. This context can hardly be overemphasized. It created an insatiable appetite for gardens, architecture and scenery that reproduced, no matter at what cost, the idealist landscape painting styles of renaissance painters like Claude Lorrain, Nicolas Poussin, and Salvador Rosa.

Whether it was natural scenery, gardens or civic design, the landscapes that appealed to the patrons of landscape arts during the eighteenth century were those which followed the rules of renaissance painting. The results of the tendency included: the miniature compositions of dwarf plant material, smaller-than-life statues, ponds, copses of trees, framing the vistas with dark verges of non-native plant material, displaying side-by-side plant and geological material not naturally found together and a visual focus on the middle ground.

The picturesque tradition, from the beginning of Joseph Addison's thesis on the subject, was a theory based on a reinterpretation of Plato's and Hume's philosophical essays on the theory of ideas. It is clear from Addison's 1712 work "The Pleasures of the Imagination"[20] that the basis of the appeal a landscape might hold was the positive association of other places, other times, and other values that it brought to mind. The picturesque aesthetic had a lot to do with thinking: the more a landscape makes the imagination work the more beautiful it is.

The legacy of the picturesque survives from its dominance in the eighteenth century to inform the contemporary popular arts of amateur photography, landscape painting, private gardening and even some components of civic design and public park ornamentation. One sees its influence in yard art — pink flamingoes, miniature human figures, cement bird baths, alpine and scree gardens, the assorted chinoiserie displayed in botanical gardens. We are reminded of the link between individualism and the picturesque by the contemporary aesthetic axiom, "Beauty is in the eye of the beholder." One can turn today, as Lowenthal and Prince turned two decades ago,[21] for reassurance about the critical importance of the picturesque to records of landscape formation in the United Kingdom during the agricultural enclosures.

Corresponding to these enclosures and the design of large, private estates were changes in scenery. At first, only sinuous Chinese-like curves grafted onto the rational plans of established estates indicated the change, but a new conception of space, and man's relation to it, was being developed. At Castle Howard, John Vanbrugh allowed the countryside, rather than formal design requirements, to serve as a basis for organizing the site. At Blenheim Palace, Thomas Bridgeman invented the *ha ha* which extended prospects and encouraged the development of landscape art which fits into the natural world. William Kent applied principles of painting to landscape art, utilizing contrasts of Italian light and shade and the serpentine line in the overall site composition.

By the mid-eighteenth century, both the attack on rationalism and the implementation of new aesthetic values in the landscape were in full swing. Both are characteristic of the landscapes created by Capability Brown who remade royal gardens and scores of parks and estate grounds for landed aristocracy. He routinely destroyed the gardens which had previously surrounded country mansions, demolishing walls, hedges, labyrinths, flower beds, parterres and ornaments in his zeal to abolish formality and evoke pastoral images.

Thousands laboured under his direction draining pools to create meadows, joining pools and damming rivers to make lakes, and excavating ground by pick, shovel, and wheelbarrow to create serpentine lines along their banks. Millions of oak, beech and ash trees were planted and transplanted in clumps on grassy meadows at strategic picture-like points. And so the picturesque landscapes of Britain were formed.

Travellers by the thousands armed themselves with texts guiding them to the proper way to "see", to respond to, scenery with picturesque terminology. One of the most influential guides to landscape tastes in the picturesque tradition was William Wordsworth's *Guide Through the Districts of the Lakes*, written in 1835.[22] At the height of picturesque influence at the beginning of the nineteenth century, the convention was so formal that English travellers used "Claude glass", gold-rose coloured prisms named after the painter Claude Lorrain's colouring of Italian landscapes, to make certain the atmospheric light of Britain looked like the atmospheric light of Italy. It was the extreme conventions of the picturesque along with developments in natural science and the economic shift from mercantile to industrial capitalism that led to a new aesthetic tradition — sublimity.

SUBLIME TRADITION

The sublime tradition, embracing the feelings of terror and human frailty generated by scenery of raw power, confronts us with a paradox. On the one hand, we are assured that this tradition is a reaction to blind conventions of the picturesque. We are assured that it is tied to scientific achievements of the geomorphologist Charles Lyell and the naturalist Charles Darwin; and, that it is the aesthetic basis for contemporary movements like the Sierra Club. On the other hand, the tradition is equally tied to Evangelical typology, the established method of symbolic interpretation of the bible, and the past and future will of God. The sublime tradition presents itself to us today as theosophy and the nature aesthetics of Zen Buddhism. John Ruskin, who more than anyone is the father of the Canadian landscape aesthetic, is the pivotal figure for geographers seeking to reconcile this dilemma.

Lowenthal and Prince,[23] amongst other geographers, have focused on the notion of sublimity. They quite correctly spoke of Edmund Burke's early distinction between the "beautiful" and "sublime" components of natural scenery but the paradox is not addressed. Burke's 1757 essay, *Philosophical Enquiry into the Origins of our Ideas of the Sublime and Beautiful*[24] was followed by Immanuel Kant's treatise in 1790, *Critique of Judgement.*[25] Together these two criticisms of landscape aesthetics created, among popular writers like Walter Scott and the Brönte sisters, an appeal for narratives set on stormy heaths, rugged coasts, majestic mountains, in dark forests and near waterfalls. The sublimity of renaissance painters like Salvador Rosa was rediscovered and Joseph Turner's impressionistic landscapes depicting natural storms and rugged topography with images that were truthful to observed nature, not idealized nature, grew in their appeal by the middle of the nineteenth century.

Ronald Rees has done a commendable job of drawing geographers' attention to the distinction between beautiful and sublime dimensions of nineteenth century landscape aesthetics.[26] But Rees did not unravel the paradox of Evangelical typology and scientific realism with the sublime tradition any more than do Lowenthal and Prince. Nor did he describe the theoretical basis for Ruskin's elaboration of the sublime and the Victorian standards of landscape aesthetics so critical to an understanding of landscape tastes in North America when the Dominion of Canada commissioned its first public buildings, its civic designs and its national park ground plans.

The paradox of the sublime aesthetic tradition is clarified once it is realized that the method of landscape aesthetics Ruskin pronounced in *Modern Painters*[27] was based on a firm distinction between allegory and typology. In allegory, for instance, in the figures of imagination produced by the picturesque, the landscape objects conveying meaning have no existence. They are fictitious devices whose function is to stand for something else. In typology, the figures are real objects, people or events which retain their identity as well as conveying a symbolic meaning. Thus for Ruskin and millions of fundamentalists, there was no doubt that the external world of natural science continued to exist and at the same time was also a direct expression of God.

Only by granting full stature to Ruskin's typology of landscape aesthetics can we make sense out of high Victorian sublimity. As was the case with the rational tradition, a major source of aesthetic feeling resulted from unity of scenery. Ruskin's unity, however, was contrasted with that of rationalism: there was a unity of subjection, origin, sequence, membership in class or of what he called cordial variety. Unity symbolized God's comprehensiveness. Hence the rhythms of waving grass and trees subjected to a common wind were found appealing, as was the common movement of clouds and ocean waves subjected to the same wind. In a sense, it was natural processes that symbolized a Divine Spirit. There was an aesthetic feeling to be experienced from visually tracing a river from its mouth to its sources. And, according to the theory of sublimity, the march of the seasons, the sequence of spring, summer, fall and winter, was another revelation of an all powerful Divine Spirit. Thus, thousands of Victorians crowded the countryside and found aesthetic pleasure in classifying buildings into stylistic periods, or trees, rocks and plants into memberships. Even if they could not classify every object, they would still be able to admire the "humanness" of God, his recognition of imperfections, of cordial variations, and of the individual. We must imagine the delight of bible-reading families finding hidden meanings in scenery the way they found hidden meanings in parables. Then there was the special significance of light on the horizon, at dawn or dusk, that was a symbol of pure divinity. The visual experience of infinity, like distant light, was also a source of special delight. In this way, architecture, civic design, vistas and gardens that offered infinite varieties of curves, variations of tone, colour or line were found appealing, especially if these infinities were somewhat self-restrained.

Notwithstanding the current revival of religious fundamentalism in North America, these elaborations of sublimity worked out by Ruskin may seem somewhat irrelevant. But, when it becomes clear that this tradition has informed the architecture of Canada's national capital, of the legislative assemblies of Ontario and British Columbia; when it is understood that the theory is the basis for the early railway stations of Canada and the decorations at entrances to national parks; that it is the ultimate source of both the Group of Seven landscape paintings and the literary work of the first national writers like Charles Roberts; then its importance to Canadian landscape tastes cannot be overstated. The scientific realism that underlay Ruskin's move from the picturesque to sublimity was further developed into a new tradition by one of his students, William Morris.

REALIST TRADITION

In describing the traditions in landscape aesthetics, there would appear to be some basis for not labelling realism as a separate tradition, a basis for including it as a dimension of other traditions. For each historical shift in the modes of producing social reality, shifts in economic, social and technical order, there are associated shifts in cultural production and aesthetics. Hence there are several "realities" like the scientific reality of the new geology and new biology that led Ruskin from the picturesque to the sublime or of quantum physics and radiation that led Gropius from variations of the picturesque and sublime to the Bauhaus aesthetic. That way of classifying aesthetic traditions would be reasonable, but for the theoretical achievements of Morris and the arts and crafts movement. This movement comprised a strong appeal for regionalism, utility and a new naturalism in the landscape arts. The principles of Morris' realism are presented in many of his essays like "Art and Socialism" (1888),[28] in his utopian novel *News from Nowhere*,[29] published in 1890, and in his earlier *Manifesto of the Society for the Protection of Ancient Buildings* (1877).[30]

Geographers ought to be especially concerned with realist aesthetics because Morris worked them out with the support and encouragement of two nineteenth century geographers — Peter Kropotkin and Elisee Reclus. Morris' natural realism had its most profound influence upon gardening, architecture and civic design. In architecture, he, his fellow Pre-Raphaelites and his followers created an appeal for building designs that were large scaled — "larger than life" or "super" real — based on the styles of local primitive building and materials rather than the styles of recognized fine art.

The larger significance of Morris' theory of realist aesthetics can be introduced through his contribution to architectural and nature conservation. Before Morris, idealist historicism ruled the day. Buildings and areas were valued as worthy of preservation only because they were the "oldest" of a kind; because they were mystically the "original" and "authentic" objects of a history. In restoring buildings and areas, the "bad" additions to them were removed and they were restored "to historical period", which commonly meant in architecture to a "better" example of the period than they were in historical reality. Morris' manifesto on conservation had a critical impact on public opinion. He wanted buildings conserved with the "good" *and* "bad" examples of taste, conserved in their matter-of-fact reality of human goodness and frailty. The public's appeal, its historical aesthetic, became increasingly realist. So today we find aesthetically pleasing both old government buildings and old factories, both old mansions and old slums.

The realist tradition in landscape aesthetics has diffused from a more realist historicism to permeate the general attitudes of the public toward scenery. The re-creation of the positive feelings we experience in sight-seeing more and more comes from matter-of-fact reality. Put simply, in our time bus drivers *do* take busman's holidays. Every geographer who takes part in the ritual of conferences and their associated field trips shares in the "delights" of visiting pulp mills, urban slums and natural marvels like canyons and waterfalls. Can that delight be much different from the delight of another time or another reality — when in the nineteenth century only the canyons and the waterfalls would have been a source of aesthetic pleasure?

If these experiences seem to be absurd, they were not meant to be. The final tradition to be examined intends the absurd, the preposterous, to be a dimension of landscape aesthetics.

SURREALIST TRADITION

Surrealism defined in its narrowest sense refers to an aesthetic movement that developed from Dadaism between World War I and II. We may use the term in a broader sense to include certain tendencies in aesthetics that predate the formal existence of surrealism between the wars and tendencies that have developed in Western art theory in recent decades. Basically, these include the tendencies to broaden the emotional response intended in landscape arts beyond those of beauty and awe to include comic relief, instinct, illusions, subconscious dream experiences, the grotesque, the absurd, and the fantastic.

So defined, the surrealist tradition has a long history. It was manifest in the appeal of the maze as a component garden in renaissance landscape design. It was also manifest in the delight experienced in the comical, surprise mechanical fountains favoured in the Baroque period. The tradition also embraced the appeal of visual illusion associated with goose-foot intersections, false facades and the *ha ha* intersection so popular in the renaissance styles of the seventeenth century and the picturesque styles of the eighteenth century.

Between World War I and II, elements of the grotesque and absurd, while never dominant, nonetheless flourished in some regions. The Art Nouveau architecture and civic design of Antonio Gaudi with imaginary animal motifs, free-line forms, and striking colours were an example. Another case was the landscape painting of Expressionists, like Vincent Van Gogh, and after World War I, the Blaue Reiter theorists whose use of strident unnatural colours and free line, non-geometric forms as expressions of human emotion and states of feeling spilled over into architecture and civic design. German and Polish architecture have notable examples of these surrealist tendencies.

In North America, where after World War II rationalism won mass acceptance in civic design and architecture, there has been a continuous appeal in popular commercial design for instinctive styles. Robert Venturi noted this appeal in his *Learning From Las Vegas*,[31] in, for example, the appeal of hot dog stands that look like hot dogs.

The opposition of the surrealist tradition to rationalism grew from a small minority of architects, planners and interior designers during the post World War II decades into a major movement — the so-called post-modern movement — during the last quarter of the twentieth century. The aesthetics of post-modernism as they relate to art in general are summarized in *The Anti-Aesthetic: Essays in Post-Modern Culture*.[32] Charles Jencks' text *The Language of Post-Modern Architecture*[33] presents the theoretical basis of the movement as it influences architecture and town planning. Both these works conclude that the principal component in post-modern landscape aesthetics is a distortion of matter-of-fact reality, a schizophrenic relationship between colours, spatial perception, historical motifs, materials, land uses and private/public being. Not surprisingly, the essays in the text by Foster claim the new absurd, surrealist aesthetic is a result of the way that electronic media — movies and television — have altered the social production of reality.

The schizophrenia of post-modern landscapes, the appeal they have for patrons and designers in our time, is to be witnessed throughout the United Kingdom and the United States. But it is nowhere more apparent in Canada than it is in Vancouver. In public squares one sees private intimate colours, pale pink or peach. Neon lights, not painted wood or stone, outline classical arches. Historical motifs are severed from their contexts and mixed in a landscape collage. Punched-out walls and deceptive painting plans give the illusion of two-dimensional not three-dimensional space. Monumental triumphant arches mimic television screens; art schools and theatres are pressed against large cement plants; and expensive townhouses flaunt the fact that they were adapted from nineteenth century factories. Perhaps the most critical change in landscape aesthetics during our time is the sequential replacement of modern rationalist by post-modern surrealistic aesthetics. But if we can learn anything from this review of the traditions of landscape aesthetics, we will understand that what is surrealistic in our time may be real in the future.

CONCLUSIONS

Ever since Vaughan Cornish there has been a broad interest in the geographical question: why is this place more appealing than that place? Can we scientifically answer this question and if so by what method should it be answered? Geographers of the Canadian West have tried to refine our ability to answer the question scientifically; certainly more has been written recently in Western Canada about science and landscape aesthetics than about history and landscape aesthetics. With this review of four centuries of the history of landscape arts, with the identification of traditions in rationalism, the picturesque, the sublime, realism and surrealism, it is obvious that in addition to scientific evidence there is historical evidence of landscape aesthetics. Geographers have described some of this historical evidence but never with a full overview of four centuries. And no doubt others will re-sort the evidence, classify it in other ways and even reinterpret it. But it is doubtful if anyone will disagree that what people have designed in the past, if taken collectively, is one of the most objective of all possible records of our aesthetic relationship to scenery. It is doubtful that anyone will be able to contest the conclusion presented here, that although tastes for scenery change in important ways there are equally important tastes that persist, that are continuous, if dialectically so. Any approach to scenery — scientific, humanistic or applied — that fails to recognize this does so at its own peril.

REFERENCES

1. This chapter was developed from a paper jointly presented with Paul Martin at the Annual Meeting of the Canadian Association of Geographers in 1984. I am indebted to Mr. Martin for writing the first drafts of the review of Lowenthal and Prince and the section on the picturesque.

2. GROPIUS, W., *The New Architecture and the Bauhaus*. London: Faber and Faber Limited, 1935.

3. CORNISH, V., *Scenery and the Sense of Sight*. Cambridge: Cambridge University Press, 1935.

4. CORNISH, V., *The Poetic Impression of Natural Scenery*. London: Sifton Praed, 1931.

5. CORNISH, V., *The Beauties of Scenery: A Geographical Survey*. London: Muller, 1943.

6. LOWENTHAL, D. and PRINCE, H., "The English Landscape", *Geographical Review*, 54, 1965, pp.309-349.

7. LOWENTHAL, D. and PRINCE, H., "English Landscape Tastes", *Geographical Review*, 55, 1965, pp. 186-222.

8. APPLETON, J., *The Experience of Landscape*. Chichester: John Wiley and Sons, 1975.

9. RELPH, E., *Rational Landscapes and Humanistic Geography*. London: Academy Editions, 1981.

10. JENCKS, C., *The Language of Post-Modern Architecture*. London: Academy Editions, 1981.

11. DEARDEN, P., "A statistical technique for the evaluation of the visual quality of landscape for land-use planning purposes", *Journal of Environmental Management*, 10, 1980, pp. 51-68; "Landscape Assessment: The Last Decade", *The Canadian Geographer*, 24, 1980, pp. 316-325 and "Landscape Assessment: The Last Decade: A Reply", *The Canadian Geographer*, 26, 1982, pp. 76-79.

12. HAMILL, L., "Quantitative Methods for Investigating the Variables that Underlie Preference for Landscape Scenes: Comment No.2" *The Canadian Geographer*, 28, 1984, pp. 286-88, and "Landscape Assessment: The Last Decade", *The Canadian Geographer*, 26, 1982, pp.76-79.

13. HAMILTON, W., "Landscape Assessment: The Last Decade: Comment No. 2", *The Canadian Geographer*, 26, 1982, pp. 73-76.

14. PEARCE, S. and WATERS, N., "Quantitative Methods for Investigating the Variables that Underlie Preference for Landscape Scenes", *The Canadian Geographer*, 27, 1983, pp. 328-344.

15. PEARCE, S. and WATERS, N., "Quantitative Methods for Investi gating the Variables that Underlie Preference for Landscape Scenes: Reply", *The Canadian Geographer*, 28, 1984, pp. 288-290.

16. This literature review is based on the following: *Research Guide to the History of Western Art* by Eugene Kleinbauer and Thomas Stevens, 1982; *Guide to the Literature of Art History* by Etta Arntzen and Robert Rainwater, 1980; and *Fine Arts: A Bibliographic Guide to Basic Reference Works, Histories and Handbooks (2nd ed.)*, by Donald Ehresmann, 1979.

17. CAMPBELL, C., *Vitruvius Britannicus*. [Reprint] New York: B. Bloom, 1967.

18. GROPIUS, *op. cit.*

19. RELPH, *op. cit.*

20. ADDISON, V., "The Pleasures of the Imagination", in *Critical Essays From the Spectator*. [Reprint, D.F. Bond (ed.)] Oxford: Oxford University Press, 1970, pp. 172-209.

21. LOWENTHAL and PRINCE, *op. cit.*

22. WORDSWORTH, W., *Guide Through the Districts of the Lakes*. 5th edition. Kendal: Hudson and Nicholson, 1835.

23. LOWENTHAL and PRINCE, *op. cit.*

24. BURKE, E., *Philosophical Enquiry into the Origins of Our Ideas of the Sublime and the Beautiful.* [Reprint, J.T. Bolton (ed.)] London: Routledge and Kegan Paul, 1958.

25. KANT, I., *Critique of Judgement.* [Reprint] Oxford: Clarendon Press, 1961.

26. REES, R., "The Taste for Mountain Scenery", *History Today*, May 1975, pp. 305-312.

27. RUSKIN, J., *Modern Painters.* [Reprint] London: J.M. Dent and Sons, Ltd., 1929-1935.

28. MORRIS, W., "Art and Socialism", cited in D.E. Egbert, *Social Radicalism and the Arts.* New York: Alfred A. Knopf, 1970, p. 449.

29. MORRIS, W., *News From Nowhere.* [Reprint] New York: Monthly Review Press, 1966.

30. MORRIS, W., *Manifesto of the Society for the Protection of Ancient Buildings.* Cited in E.P. Thompson, *William Morris.* London: Merlin Press, 1977, p. 228.

31. VENTURI, R., *Learning From Las Vegas.* Cambridge: M.I.T. Press, 1977.

32. FOSTER, H., (ed.), *The Anti-Aesthetic: Essays in Post-Modern Culture.* Port Townsend: Bay Press, 1984.

33. JENCKS, C., *op. cit.*

PLATE 7 Pastoral scenes, such as these cows grazing in the Fraser Valley of British Columbia, are very popular in many cultures. ►

PLATE 8 Baie de Verde in Newfoundland has many elements of a traditional subsistence landscape. Such scenes may often appear attractive to outsiders but mask an existence perilously close to the poverty level. ►

3 SOCIETAL LANDSCAPE PREFERENCES: A PYRAMID OF INFLUENCES

Philip Dearden

INTRODUCTION

Different philosophical bases underlie various methodological approaches in landscape aesthetics. An objectivist stand predisposes methodology toward concentration on landscape — beauty is inherent in objects — and the goal, therefore, is to assess these objects; a subjectivist stand leans heavily towards examination of individual preferences, beauty being in the eye of the beholder.[1] Examples of early approaches within these modes would include Linton[2] and Leopold[3] in the objectivist mode, and works by the Kaplans[4] and papers in the book edited by Gold and Burgess[5] in a more subjective mode. However, much current research adopts a more relational view that landscape preference is a combination of the two viewpoints that must, of necessity, involve elements of both object and subject. Psychophysical techniques such as those by Dearden[6] and Schroeder and Daniel[7] exemplify this approach.

An important methodological consideration underlying all approaches is the degree of consensus on landscape preferences within the population. Some studies have revealed an astonishingly high degree of consensus[8] whereas others have plotted distinct differences.[9] Both may be correct. Overall, consensus appears high but certain landscapes and certain populations can interact to produce significant differences. Such differences could be related to landscape characteristics, observer characteristics or mode of interaction between the two. The purpose of this chapter is to concentrate on observer characteristics suggested in the literature that appear to influence landscape preferences. This is of interest not only theoretically but also in terms of methodology in ascertaining which variables are methodologically relevant for landscape assessment techniques. The paper draws together findings related to between-group differences in landscape evaluations and suggests an organizational framework described in the following section.

THE FRAMEWORK

Many observer-dependent factors have been suggested as influencing landscape preferences. The most commonly recognized and researched are shown in Figure 1,3. Several points should be made:

1. Although the diagram presents the variables as discrete they are not so and may interact in numerous different ways.
2. It is often very difficult methodologically to adequately discriminate between these variables in terms of influence on landscape preferences.
3. Although the diagram presents the variables in a pyramidal hierarchy this is not meant to imply the relative importance of the variables in determining any given landscape preference. It is intended to indicate that all these factors are present, with the hierarchy reflecting more the potential degree of societal consensus related to each variable rather than relative influence of each on any given individual. This is shown on the left hand side of the pyramid where it is suggested that innate factors, conditioned by our evolutionary history, are common to mankind. Cultural differences in taste for landscapes may be common to that particular society, whereas familiarity would have regional importance and various socio-economic differences vary from individual to individual. Hence the degree of potential individual differences rises with the hierarchy.
4. The above observation has methodological implications. Techniques that are dominantly landscape-based would be more appropriate to use in landscape assessment studies where individual differences would be expected to have a relatively minor influence compared to similarities generated by innate and cultural factors. However the necessity for more in-depth individual probing and use of observer-based assessment techniques rises as the potential degree of individual differences rises through the pyramid. This is shown on the right hand side of the framework.
5. The sum total of these influences is societal landscape preferences — the variable that forms a crucial part of many landscape assessment techniques. The degree of consensus within societal landscape preferences is seen as a reflection of the composition and relative influence of the variables in the pyramid in conjunction with landscape characteristics. Discussion will now turn to explore the available literature on the influence of each variable in the pyramid.

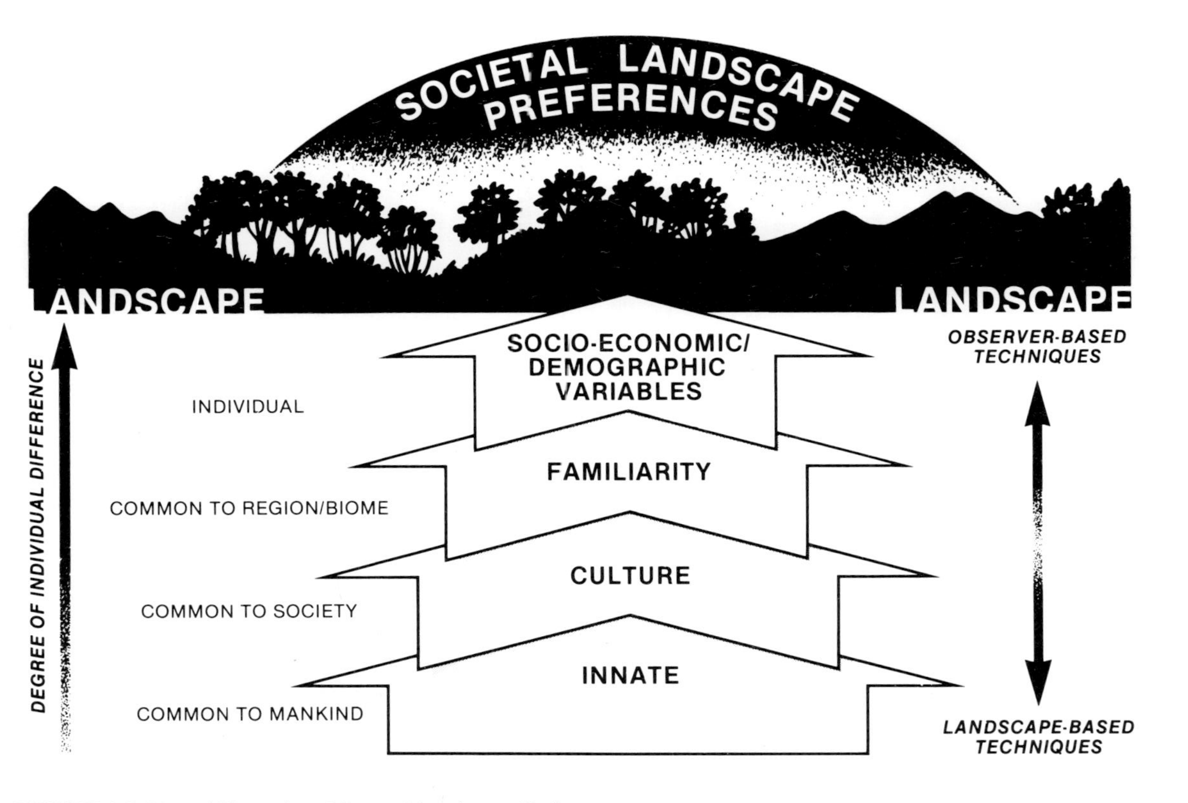

FIGURE 1,3 Nested Hierarchy of Societal Landscape Preferences

INFLUENCES ON SOCIETAL PREFERENCES

Innate

At the base of the theoretical framework of influences on societal preferences are those factors common to mankind as a reflection of the type of biological organism we are. Perhaps most obvious of these is the five senses — sight, smell, hearing, touch and taste that constitute our information-gathering apparatus from the environment. Although the efficacy of these differ from individual to individual the similarities are more striking than the differences. We have collectively poorly developed olfactory data collecting capability compared to dogs for example, but are blessed with stereoscopic, binocular, colour vision, and collect most of our information from the outside world by this means. And this gives us a good degree of commonality. It would be difficult indeed to imagine any sort of consensus on landscape emerging if we all collected our environmental information by radically different means.

Not only do we collect our information using similar apparatus, our processing capabilities for that data also have a great deal in common. Our common evolutionary history, it is suggested by various authors, predisposes us to prefer certain types of landscape. Appleton,[10] for example, suggests that landscapes offering both prospect and refuge, the ability to see without being seen, are landscapes favourable to the biological requirements of humans. This kind of proposition is exceedingly difficult to test. In a somewhat limited attempt with only four subjects Clamp and Powell[11] failed to provide empirical support.

A related attempt was made by Balling and Falk[12] who, on the basis of work by Appleton and others, hypothesized that there exists an innate preference for savanna-type landscapes (that would provide both prospect and refuge) that is modified over time by environmental experience. Using slides of five different biomes — tropical rain forest, temperate deciduous forest, coniferous forest, savanna and desert — with a wide selection of age groups they found that elementary school children showed a significant preference for savanna over all other biomes. Other natural environments became equally preferred with older subjects. From this they conclude that there is some support for the hypothesis that there is "some innate preference for savanna-type environments, arising from the long evolutionary history of humans on the savannas."[13]

At a more general level several authors[14] have suggested that evolutionary factors are also responsible for the almost universal preference for natural over built environments. Humans, it is argued, are innately predisposed to respond more favourably to natural settings through biological conditioning. Not only are natural settings preferred, levels of consensus are much higher for natural than built settings.[15] It may be suggested that with reference to Figure 1,3, innate factors are perhaps more important in determining societal landscape preferences for natural landscapes, whereas variables further up in the hierarchy become more important with built environments. It should also be pointed out that natural settings are not necessarily wilderness. Ulrich maintains that "American groups appear to respond to a scene as natural if (1) it contains extensive vegetation or water, and (2) if buildings, cars and other built features are absent or not prominent."[16] With regard to the first point it should be noted that there is a large amount of both intuitive and empirical literature indicating the universal appeal of water as a landscape element.[17] Ulrich suggests that this may also be a biologically based response.[18] Humans have low water holding capabilities within their bodies and have evolved to place a high value on the presence of water. A common tangible reflection of this is the high value associated with homes with water views. People are willing to incur a substantial economic cost to gain an intangible psychological advantage.

Water and natural elements are content variables. Kaplan and Kaplan[19] have developed a theoretical framework that concentrates on process as opposed to strict content. Process "refers to patterns that are content-free or applicable across different content domains."[20] Landscapes that are likely to be preferred, they argue, are those that permit 'involvement' and 'making sense'. Involvement is enhanced in complex environments with mystery, making sense is facilitated by coherence and legibility, "characteristics that permit one to interpret readily what is going on and that facilitates seeing where one is headed."[21] The Kaplans suggest that these information processing capabilities are the result of our evolutionary history. There was survival value placed on the ability to quickly assess and react to situations. As a result, man developed the ability to build cognitive maps to enable him to interpret the environment. These maps and our abilities to create them have evolved over millions of years and are still with us today, influencing our landscape preferences.

There appears to be no conclusive evidence from the literature that would empirically support a cause and effect relationship between evolutionary history and landscape preferences. Intuitively, however it appears unlikely that the ideas proposed by authors such as the Kaplans, Appleton, and Balling and Falk would not have some influence on landscape preferences, albeit one that tends to emphasize commonality between groups and individuals rather than differences. These perceptions underlying human preferences would hence form a common base upon which the subsequent layers in the hierarchy (Figure 1,3) would be superimposed. Lyons[22] has argued for the primacy of these subsequent layers over any innate, heritable component of landscape preference. She criticizes the age-related evolutionary hypothesis put forward by Balling and Falk discussed earlier and suggests that it is critical for the hypothesis to be supported:

> that a preference be found for East African savanna and not for the savanna-like parks and backyards in which young children spend much of their time. Without such evidence, evolutionary explanations of landscape preference appear less parsimonious that do those that do not presuppose innate preferences and yet account for a significant amount of preference variation with socially differentiating factors.[23]

If the goal is to determine preference variation between groups and individuals then the available evidence suggests that 'socially-differentiating' factors provide fairly good predictive power. However this is not to claim that innate factors do not constitute a significant influence; they simply have a much wider degree of commonality in humans derived from our common ancestry.

Superimposed upon this common tapestry are various "tastes". As Jay Appleton explains:

> All cats, unlike all rabbits, prefer to assuage their hunger by eating flesh, but some cats grow up with a preference for meat, others for fish. All kinds of facts and fancies may distinguish one individual cat from another, yet it is impossible to avoid the conclusion that there is a common range of preferences which can be observed in all but those most aberrant of cats. These characteristics, physical and behavioural, are inborn; but the behavioural, and in some cases even the physical characteristics may be developed or modified along idiosyncratic lines in each individual as a result of the sum of its experiences. In other words we develop our own preferred methods of gratifying common, inborn desires. We acquire 'tastes'.[24]

Perhaps the most encompassing of such tastes are those conditioned by particular cultures. The next section will discuss some of the research that has been directed toward the influence of culture on landscape preferences.

Culture

As an organism we may react to environment in ways distinguished more by their similarity than differences. However, the environment is not uniform. Different cultures perceive, adapt to, and alter the environment in many different ways. Indeed early environmental determinists such as Huntington[25] and Semple[26] offered environment as the major determining factor in cultural differences. Views are not now so rigid nor so unidirectionally linear. Environment influences culture, and vice versa, as expressed by Lowenthal and Prince in their article on English Landscape tastes:

> Landscapes are formed by landscape tastes. People in any country see their terrain through preferred and accustomed spectacles, and tend to make it over as they see it. The English landscape, as much as any other, mirrors a long succession of such idealised images and visual prejudices.[27]

They go on to outline six such visual prejudices: the bucolic, the picturesque, the tidy, facadism, the past and *genus loci*, and point out, "Few of these traits are in fact exclusively English; they are distinctive, however, in the sense that they occur throughout the realm but not generally elsewhere".[28]

In a later article[29] Lowenthal looks at the American landscape and suggests some of its characteristics, some innate, others culturally determined; size, wildness, formlessness, extremes, the future, the past, featurism and glorification of the remote. Walter[30] in a more recent essay has offered his insight into some of the cultural differences in landscape perceptions between Americans and English following his first visit to the American continent. As an Englishman he was confused, and generally uncomplementary about American landscapes, "...once you have learned one language, it is strange to be in a country that also speaks English but speaks a different visual language. It's as though the English landscape, like English culture, speaks in subtle shades, while the American landscape and culture speak in primary colors."[31]

One of the major differences between the cultures, he maintains, is simply the attention given to the "Great Outdoors" in American culture. "The great landscapes of the West are not presented as scenery, as something aesthetic to be enjoyed by those who are interested and passed over by those who are not; rather they are treated as archetypically American, to be revered and venerated by all."[32] He suggests that other cultures have similar disparities in the place of nature in the national consciousness. Although France, Italy and Germany are all Alpine nations, national images depicted on travel posters rarely show alpine aspects of the first two, while that is perhaps the predominant image of Germany, healthy and outdoors, despite the fact that it is no more mountainous nor spectacular than France or Italy.

Secondly, the American landscape has a different kind of beauty. Walter's main destination was the Colorado Rockies — some of the most spectacular scenery on the continent, but he remains unimpressed, firmly nurtured in his English traditions. The mountains, though high, have no glaciers. The sun is too high, giving a visual flatness to even quite dramatic landscapes, compared to the soft, low light of Britain. The sun shines too much, giving too little visual contrast and change. There are too many trees. They are all the same. They block the view. The English dislike for trees, he links to differences in English and American cities. The latter are newer with large lots and fewer trees. Americans do not lack openness in their urban environments, he suggests, in contrast to the more claustrophobic English cities — America's forests, "...provide securities and reassurance for a society that is open geographically as well as culturally."[33]

Perhaps the chief difference he discerns in cultural tastes between Europeans and Americans is, "...the difference between beauty based on essence and beauty based on balance."[34] European landscapes are complex, a mixture of elements. Their beauty stems from the way these elements have been woven together over centuries. By way of contrast the finest American landscapes are not a sensitive balance of many elements, but the pure essence of one — desert, mountain, canyon or forest. Urban gardens mirror this landscape, he suggests. European gardens are a mixture of flower garden, lawn, vegetable garden, trees, usually walled or hedged in. North American garden is the lawn — a vast expanse of unfenced grass — pure element, in front of the home for all to see.

This is not to say that the American landscape does not have such intricate balances. Walter finds Yosemite Valley particularly pleasing with its mixture of cliff, forest, meadow, river, dome and waterfall. More Arcadian landscapes also exist elsewhere in North America — New England springs to mind, but, it must be admitted, these are not landscapes representative of the continent. So, Walter concludes:

> Nature may reign supreme in the courts of beauty in America, but culture put her there. If you do not share the culture, the queen looks odd. Americans often asked me to explain our royal family because they couldn't understand something so strange. I now realise that asking Denverites about their reverence for the Rockies is not so different.[35]

These personal observations lend some insight on the influence of cultural glasses on landscape. It should also be remembered that many Americans were at one time Europeans and considerable literature has been devoted toward discussing the transformations in landscape tastes over time by various cultures.[36] Other authors, such as Tuan,[37] have also documented differences in perception of the same North American landscape by indigenous and immigrant people.

Much of this literature is descriptive. The experimental evidence lends much less support to the importance of cultural differences in landscape perceptions and preferences, especially between relatively similar landscapes and cultures.[38] Differences that have been documented tend to be between very different cultures[39] or involving quite specific environmental settings, particularly urban,[40] or related to specific environmental attributes.[41] Psychologists, such as Cole and Scribner,[42] have also pointed to a lack of sound experimental evidence for culture differences in fundamental perceptual and cognitive processes. Hence, from this brief review, it seems that the descriptive literature on the influence of culture on aesthetics is rich, but the experimental evidence, sparse. Clearly more research is required in the area to more fully understand this apparent paradox.

Familiarity

"Familiarity breeds affection when it does not breed contempt" asserts Tuan in his well-known book *Topophilia* and there is a wealth of research data to support him. Early work by Lynch[44] on mental images of cities found familiarity to be important as did later urban work by Appleyard.[45] Several authors have found familiarity of respondents either with the actual sites under evaluation or similar sites to be a major factor influencing landscape preferences.[46] Clamp,[47] amongst some other authors, has found no relationship between home region of respondents and landscape preferences yet feels that, "it is a matter of everyday experience that given a choice between the familiar and the unfamiliar, even with a few extra incentives thrown in, most people will prefer the familiar."[48] He goes on to suggest that, "it may well be that preference for familiar landscape is restricted to those more local places known specifically to individuals, and to which adhere particular knowledge and particular memories."

This brings into the question the problem of scale. Respondents could be potentially familiar with a specific site (as suggested by Clamp above), with regional characteristics but not the specific site, or a more generic familiarity with the landscape type. Several authors report research related to the latter aspect. Lyons,[49] for example in a study of biome preference, presents findings that suggest that landscape preference is strongly influenced by residential experience in different biomes. Dearden[50] adds support not in terms of biomes but general familiarity with certain types of landscape. He found, for example, a positive correlation between housing density occupied most of adult life and preferences, with respondents from low density, dominantly natural housing environments expressing higher appreciation of more natural scenes and vice-versa. He also reports that those with most wilderness contact have higher ratings for wilderness scenes. In an attempt to elicit which factors respondents themselves thought exerted most influence on their preferences he found that the top four mentioned, past landscape experience, travel, present living environment and recreation activities all seemed to be closely related to familiarity.

There is less support for regional familiarity as an influence on preferences. Daniel and Boster[51] found, for example, very similar preferences expressed by subjects irrespective of origin. Wellman and Buhyoff[52] undertook a study specifically directed toward the effects of regional familiarity on landscape preferences and found no regional familiarity effect. Shafer's experiments to replicate his American model elsewhere[53] also failed to reveal marked regional effects on landscape preferences, as did Clamp's work in England.[54]

In terms of specific familiarity with scenes the evidence suggests potential for a much closer relationship. Although some studies[55] have revealed negative associations others, such as those by Jackson *et al.*[56] and Nieman[57] have found evidence for a positive correlation. Herzog *et al.*[58] found familiarity to be a very effective predictor variable in preferences for urban scenes. They suggest that in reacting to a familiar scene the subject is not reacting to the stimulus *per se* but to a whole set of experiences and knowledge connected to the depicted scene. Hammitt[59] specifically tested familiarity in the preference component of on-site recreational experiences. The influence of familiarity on visitor preference was measured in terms of degree of visitor preference for visual scenes as a result of acquaintance gained through three means of familiarity: on site-experiences, viewing of photographic materials prior to an on-site experience and number of previous visits. He found that both prior visits and on-site experiences increased preference levels.

Undoubtedly the relationship between familiarity and preference is not a simple one. Kaplan and Kaplan[60] give some reasons as to why this might be so and provide the matrix shown in Figure 2,3. Involvement and making sense are, they maintain, simultaneous needs that are closely related to preference. Unfortunately these requirements do not always pull in the same direction and familiarity is one variable where this might particularly be so. Life, they suggest, is a continuous trade-off between the "excitement of the new and comfort of the known."[61] On the one hand humans need to make sense of the environment quickly and efficiently. Hence the attraction of familiarity, we have the necessary cognitive maps to deal with the situation. However, people also feel a need to make use of their mental capacities — they need to be involved — or else

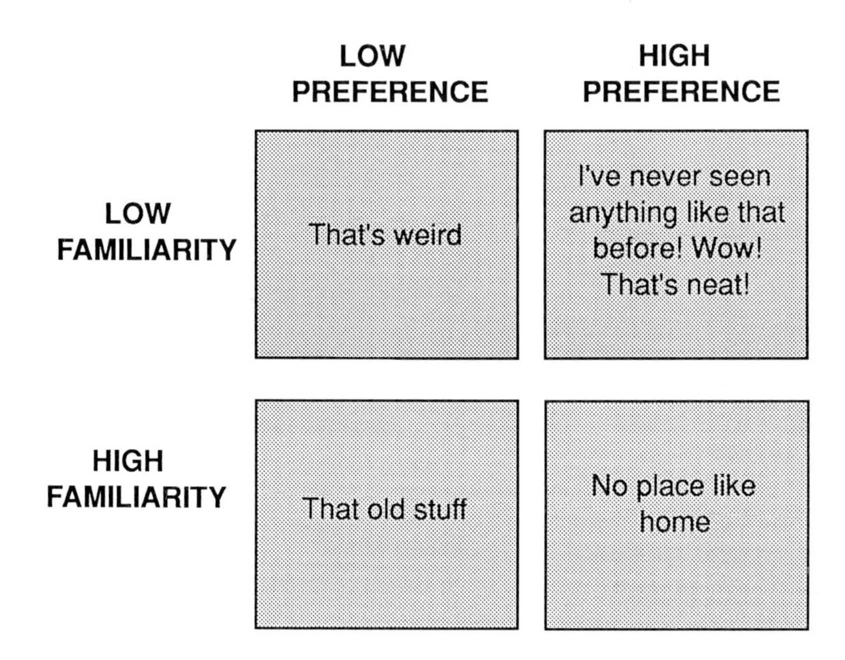

FIGURE 2,3 Familiarity and Preference Matrix Suggested by Kaplan and Kaplan (1982)

they get bored, and this is best stimulated in unfamiliar environments. A paradox develops — familiarity may both help and hinder preference. Williams[62] has provided some empirical data to support a non-monontonic view of the familiarity — preference relationship.

Preferences may change over time as a result of familiarity. Preferences may fall, for example, as familiarity increases due to decreasing necessity for involvement — this is the familiarity breeds contempt role. On the other hand, we may become more comfortable with the particular situation and preferences rise with familiarity because of our efficient cognitive maps that have developed. We know what to do. The Kaplans argue that involvement can be sustained in a familiar scene by a shift from exploration to play:

> Familiarity, the outcome of exploration, is also the starting point for play. In play the focus is not "What is this?" but "what can I do with this?"... With increasing familiarity, the mental entities become increasingly compact, increasingly discrete, and increasingly responsive to activation in the absence of what they represent ... Thus familiarity is essential to the playful rearrangement and recombination of the elements of thought that we tend to associate with insight and creativity.[63]

Increasing numbers of studies are emerging on preference-familiarity and are teasing out some consensus on the relationship. Familiarity it seems, in its various guises, has the potential to be one of the most influential of variables affecting landscape preferences.

Socio-Economic

A large number of studies suggest that various socio-economic and demographic factors act differentially on the population to influence different landscape perceptions and preferences. Subjects may have common evolutionary and cultural backgrounds, even a core of commonality with familiar landscape types, regions or scenes, but a whole multitude of personal variables related to individual experience and station in life interact to produce a seemingly endless array of landscape preferences. These variables are not independent of each other and in some cases aggregates, such as social class, have been used to differentiate between groups. Duncan,[64] for example, found that landscape tastes correlated with social class. However several socio-economic and demographic variables have been shown to be highly associated with landscape preferences.

Age has surfaced as one variable of influence in several studies. Zube, Pitt and Evans,[65] for example found that scenic values change over the lifespan. Both children and the elderly are less likely to view signs of human intervention in the landscape as detrimental compared to other age groups. They further found that high naturalism was relatively unimportant to young children. Research by Holcomb[66] supports this view. Lyons[67] found age, either as a single variable, or combined with gender to be her most powerful predictor variable discriminating between groups in a study of biome differences. Balling and Falk[68] also report age-related preference differences as does Penning-Rowsell,[69] but there are also some experimental studies that have not found age to be a significant factor.

Most of the experimental work cited above assumes that age-related differences in the sample result from the aging process of individuals rather than the possibility of different landscape values between generations. In other words, children will have similar preferences to older age groups when they get to that age and the latter, at one time, had similar preferences to current children. Little work has been done to validate this hypothesis. If found incorrect, the findings showing the new generations' greater tolerance for man-dominated landscapes has interesting implications.

Most studies assume, however, that as subjects age their values and preferences are constantly modified by their total environmental context. Balling and Falk[70] propose that their observation of an age-related decrease in preferences for savanna-type landscapes may be due to an innate evolutionary preference for such landscapes that is dominant as a child but is culturally modified with age. Lyons[71] disagrees feeling that evidence for any innate, heritable component has yet to be demonstrated.

In addition to the experimental work on the influence of age, more descriptive support comes from authors such as Tuan[72] and powerful evocations of childhood landscapes:

> Edith Cobb has said that the 'embryology of mind' requires landscape. What can be said of the mind's later growth, when landscapes become remembered landscapes? The image of youth as a last paradise is ancient and widespread. Resplendent with recollections of golden moments of leisure, the images are inextricably connected with places that give them significance.[73]

More attention is also being devoted to the elderly and their perceptions and values of space or place. A phenomenological inquiry by Rowles[74] looks in-depth at the space-place relations of five elderly people and how they have changed over time. This kind of methodological, longitudinal approach can add greatly to our understanding of age as a discriminatory variable in landscape preferences.

As age increases, more discriminatory variables appear to influence preferences. Macia[75] found significant differences due to gender and Dearden[76] reports differences at the $p < 0.05$ level for males and females in evaluations of peri-urban environments, with females showing higher appreciation of such scenes. Peterson[77] also found black females to have stronger preferences for urban scenes than whites. Lyons[78] found evidence that gender differences in landscape preferences appear to increase with age up to college-age and Tuan assures us that male and female are not arbitrary distinctions: "...Physiological differences between man and woman are clearly specifiable, and the differences can be expected to affect their ways of responding to the world."[79] Certainly the wider geographical literature on the topic, such as mental map structure, would tend to support this, but there is still ambiguity in the experimental literature as to when such differences occur and how they are manifest geographically, let alone whether gender is a "pure" explanatory variable rather than a surrogate for a wide variety of socializing influences.

One of the more studied personal variables is the influence of occupation, particularly as it relates to training in some sort of landscape-based profession. Some studies reveal quite distinct differences between experts in the field, such as architects and planners,[80] whilst others fail to reveal any.[81] Kaplan and Kaplan[82] suggest that such experts can be both strength and weakness in evaluating environmental problems or assessing landscape quality. Experts are experts because they have efficient cognitive frameworks that help them to immediately assess and diagnose problems. Such frameworks are a result, largely, of their professional training. This can be a strength in terms of problem-solving in some areas but not necessarily in others. Experts can become very set in their ways and not open to new ways of collecting and manipulating information. Landscape evidence[83] shows that like-professionals, range managers, forest managers, and planners, for example, have very high within-group consensus (they are all using similar learned generic maps) but low between-group consensus. This is obviously of some concern in the field of landscape aesthetics since such professionals often constitute at least part of landscape evaluation techniques.[84] Dearden[85] has discussed this question of professional versus public input into landscape assessment studies.

In addition to the personal variables discussed here there are a whole host of findings on various other variables and combinations. It is, however, difficult to find unambiguous conclusions on any one. Most have some studies providing evidence for their discriminating abilities and others the opposite. This could be because, as Ulrich suggests, the differences are so minor that studies with slightly different designs might yield contradictory results:

> the relative lack of emphasis in environmental aesthetics on individual variability is perhaps understandable in light of the high levels of agreement among observers reported by many investigators. Future studies should systematically evaluate individual differences along environmentally relevant dimensions of personality, rather than exclusively in terms of traditional demographic variables such as age, sex and occupation.[86]

However, to date little research has addressed the relationship between variables such as personality or values and preferences.[87]

CONCLUSION

The influence of the variables discussed above and shown in the hierarchy in Figure 1,3 interact with various landscapes to produce societal landscape preferences. At times the influence of the actual landscape under assessment, its biophysical components and their interactions, may constitute the dominant influence of societal landscape preferences. With other types of landscape the influence of component variables might be quite minor compared to the subject-related variables discussed here. Where this balance lies can be quite significant in terms of determining appropriate methodologies of approach. Why adopt a landscape-based assessment technique if major variability in preferences is hypothesized to be generated more by subject-related variables? The foregoing discussion offers a conceptual framework of organization for such subject-related variables. This same hierarchy of forces can be viewed as variables influencing societal preferences or disaggregated to the individual subject level as veneers of influence on personal landscape preferences. Whichever way it is taken it is clear that much research remains to be done to clarify the roles of various subject-related variables on landscape preferences.

REFERENCES

1. See DEARDEN, P., "Consensus and a theoretical framework for landscape evaluation", *Journal of Environmental Management*, 24, 1987, pp. 267-278.

2. LINTON, D.L., "The assessment of scenery as a natural resource", *Scottish Geographical Journal*, 84, 1968, pp. 218-238.

3. LEOPOLD, L.B., "Landscape Aesthetics", *Natural History*, 78, 1969, pp. 36-45

4. For example see KAPLAN, S., KAPLAN, R., and WENDT. J.S., "Rated preference and complexity for natural and urban visual material", *Perception and Psychophysics*, 12, 1972, pp. 354-356.

5. GOLD, J.R. and BURGESS, J., *Valued Environments*. London: George Allen and Unwin, 1982.

6. DEARDEN, P. "A statistical technique for the evaluation of the visual quality of the landscape for land-use planning purposes", *Journal of Environmental Management*, 10, 1980, pp. 51-68.

7. SCHROEDER, H.W. and DANIEL, T.C. "Progress in predicting the perceived scenic beauty of forest landscapes", *Forest Science*, 27, 1981, pp. 71-80.

8. For example COUGHLIN, R.E. and GOLDSTEIN, K.A., "The extent of agreement among observers on environmental attractiveness", Regional Science Research Institute Discussion Paper Series, No. 37, Philadelphia: Regional Science Research Institute, 1970; ZUBE, E.H., PITT, D.G. and ANDERSON, T.W., "Perception and prediction of scenic values of the Northeast", in ZUBE, E.H., BRUSH, R.O., and FABOS, J.G. (eds.), *Landscape Assessment: Values, Perceptions and Resources*. Stroudsburg, Pa: Dowden, Hutchinson and Ross, 1975, pp. 151-167; PENNING-ROWSELL, E.C., "The social value of English landscapes", in ELSNER, G.H. and SMARDON, R.C. (eds.), *Our National Landscape*. USDA, Pacific Southwest Forest and Range Experiment Station, Berkeley, California, 1979, pp. 249-255.

9. For example BRUSH, R.O., "Perceived quality of scenic and recreational environments", in CRAIK, K.H. and ZUBE, E.H. (eds.), *Perceiving Environmental Quality*. New York: Plenum Press, 1976, pp. 47-58; BUHYOFF, G.J., WELLMAN, J.O., HARVEY, H., and FRASER, R.A., "Landscape architects' interpretations of people's landscape preferences", *Journal of Environmental Management*, 6, 1978, pp. 255-262.

10. APPLETON, J., *The Experience of Landscape*, London: Wiley, 1975.

11. CLAMP, P. and POWELL, M., "Prospect-refuge theory under test", *Landscape Research*, 7, 1982, pp. 7-8.

12. BALLING, J.D. and FALK, J.H., "Development of visual preference for natural environments", *Environment and Behavior*, 14, 1982, pp. 5-28.

13. *Ibid.*, p.25.

14. KAPLAN, S. and KAPLAN, R. *Cognition and Environment*. New York: Praeger, 1982; ULRICH, R.S., "Aesthetic and affective response to natural environment", in ALTMAN, I. and WOHLWILL, J.F. (eds.), *Behavior and the Natural Environment*. New York: Plenum Press, 1983, pp 85-125.

15. COUGHLIN, R.E. and GOLDSTEIN, K.A., *op. cit.*; ZUBE, E.H. *et al.*, *op. cit.* (reference 8).

16. ULRICH, R.S., *op. cit.*, (reference 14), p. 111.

17. For example TUAN, Y-F., *Topophilia*. Englewood Cliffs, New Jersey: Prentice Hall, 1974; ZUBE, E.H., PITT, D.G. and ANDERSON, T.W., *op. cit.*, (reference 8); PENNING-ROWSELL, E.C., *op. cit.*, (reference 8); ULRICH, R.S., "Natural versus urban scenes: some psychophysiological effects", *Environment and Behavior*, 13, 1981, pp. 523-556; PURCELL, A.T. and LAMB, R.J., "Landscape perception: an examination and empirical investigation of two central issues in the area", *Journal of Environmental Management*, 19, 1984, pp. 31-63.

18. ULRICH, R.S., *op. cit.*, (reference 14).

19. KAPLAN, R. and KAPLAN, S., *op. cit.*, (reference 14).

20. KAPLAN, R., "The role of nature in the urban context", in ALTMAN, I. and WOHLWILL, J.F. (eds.), *Behavior and Natural Environments*. New York: Plenum Press, 1983, pp. 127-161.

21. *Ibid.*, p. 132.

22. LYONS, E., "Demographic correlates of landscape preference", *Environment and Behavior*, 15, 1983, pp. 437-511.

23. *Ibid.*, p. 507.

24. APPLETON, J., "Pleasure and the perception of habitat: a conceptual framework", in SADLER, B. and CARLSON, A. (eds.), *Environmental Aesthetics*. Western Geographical Series Vol. 20, Victoria, B.C.: University of Victoria, 1982, p. 31.

25. HUNTINGTON, E., *The Character of Races as Influenced by Physical Environment, Natural Selection and Historical Development*. New York: Charles Scribners Sons, 1924.

26. SEMPLE, E.C., *Influences of Geographic Environment on the Basis of Ratzel's System of Anthropo-Geography*. New York: Henry Holt, 1911.

27. LOWENTHAL, D. and PRINCE, H., "English landscape tastes", *Geographical Review*, 55, 1965, pp. 186-222.

28. *Ibid.*

29. LOWENTHAL, D., "The American Scene", *Geographical Review*, 58, 1968, pp. 61-68.

30. WALTER, J.A., "You'll love the Rockies", *Landscape*, 27, 1983, pp. 43-47.

31. *Ibid.*, p. 47.

32. *Ibid.*, p. 43.

33. *Ibid.*, p. 45.

34. *Ibid.*, p. 46.

35. *Ibid.*, p. 47.

36. HONOUR, H., *The New Golden Land*. New York: Pantheon Books, 1975; REES, R., "The Scenery Cult: Changing Landscape Tastes Over Three Centuries", *Landscape*, 19, 1975, pp. 39-46.

37. TUAN, Y-F., *op. cit.*, (reference 17).

38. SHAFER, E.H. and TOOBY, M., "Landscape preferences, an international replication", *Journal of Leisure Research*, 5, 1973, pp. 60-65; ZUBE, E.H. and MILLS, C.V., "Cross-cultural explorations in landscape perception", in ZUBE, E.H. (ed.), *Studies in Landscape Perception*. Amherst, Mass.: University of Mass., 1976, pp. 162-169; ULRICH, R.S., "Visual landscape preference: a model and application", *Man-Environment Systems*, 7, 1977, pp. 279-293. KAPLAN, R. and HERBERT, E.J., "Cultural and sub-cultural comparisons in preferences for natural settings", *Landscape and Urban Planning*, 14, 1987, pp. 281-293.

39. SONNENFELD, J., "Environmental perception and adaptation level in the Arctic", in LOWENTHAL, D. (ed.), *Environmental Perception and Behavior*. Chicago: University of Chicago, 1967, pp. 42-59.

40. ZUBE, E.H. and PITT, P.G., "Cross-cultural perceptions of scenic and heritage landscapes", *Landscape Planning*, 8, 1981, pp. 69-88.

41. ZUBE, E.H., VINING, J., LAW, C.S., and BECHREL, R.B., "Perceived urban residential quality: A cross-cultural bimodal study", *Environment and Behavior*, 17, 1985, pp. 327-350.

42. COLE, M. and SCRIBNER, S., *Culture: A Psychological Introduction*. New York: Wiley, 1974.

43. TUAN, Y-F., *op. cit.*, (reference 17), p. 99.

44. LYNCH, K., *The Image of the City*. Cambridge, Mass.: MIT Press, 1960.

45. APPLEYARD, D., "Why buildings are known", *Environment and Behavior*, 1, 1969, pp. 131-156.

46. BECKETT, P.H.T., "Interaction between knowledge and aesthetic appreciation", *Landscape Research*, 1, 1974, pp. 5-7; JACKSON, R.H., HUDMAN, L.E., and ENGLAND, J.L., "Assessment of the environmental impact of high voltage power transmission lines", *Journal of Environmental Management*, 6, 1978, pp. 153-170; NIEMAN, T.J., "The visual environment of the New York coastal zone: user preferences and perceptions", *Coastal Zone Management Journal*, 8, 1980, pp. 45-61.

47. CLAMP, P., "The landscape evaluation controversy", *Landscape Research*, 6, 1981, pp. 13-15.

48. *Ibid.*, p. 15.

49. LYONS, E., *op. cit.*, (reference 22).

50. DEARDEN, P., "Factors influencing landscape preferences: an empirical investigation", *Landscape Planning*, 11, 1984, pp. 293-306.

51. DANIEL, T.C. and BOSTER, R.S., *Measuring Landscape Aesthetics: The SBE Method*. Forest Service Research Paper RM-67, Fort Collins: Colorado, Rocky Mountain Forest and Range Experiment Station, 1976.

52. WELLMAN, J.D. and BUHYOFF, G.J., "Effects of regional familiarity on landscape preferences", *Journal of Environmental Management*, 11, 1980, pp. 105-110.

53. SHAFER, E.L. and TOOBY, M., *op. cit.*, (reference 38).

54. CLAMP, P. "Evaluating English landscapes — some recent developments", *Environment and Planning A*, 8, 1976, pp. 79-92.

55. PENNING-ROWSELL, E.C., GULLETT, G.H., SEARLE, G.H. and WITHAM, S.A., *Public Evaluation of Landscape Quality*. Planning Research Group Report 13, Middlesex Polytechnic, Enfield, England, 1977; DEARINGER, J.A., "Measuring preferences for natural landscapes", *Journal of the Urban Planning and Development Division*, 105, 1979, pp. 63-80.

56. JACKSON, R.H., *op. cit.*, (reference 46).

57. NIEMAN, T.J., *op. cit.*, (reference 46).

58. HERZOG, T.R., KAPLAN, S., and KAPLAN, R., "The prediction of preference for familiar urban places", *Environment and Behavior*, 8, 1976, pp. 627-645.

59. HAMMITT, W.E., "The familiarity — preference component of on-site recreational experience", *Leisure Sciences*, 4, 1981, pp. 177-193.

60. KAPLAN, S. and KAPLAN, R., *op. cit.*, (reference 14) .

61. *Ibid.*, p. 79.

62. WILLIAMS, S., "How the familiarity of a landscape affects appreciation of it", *Journal of Environmental Management*, 21, 1985, pp. 63-67.

63. KAPLAN, S. and KAPLAN, R., *op. cit.*, (reference 14), p. 93.

64. DUNCAN, J.S., "Landscape taste as a symbol of group identity: Westchester County Village", *Geographical Review*, 63, 1973, pp. 334-343.

65. ZUBE, E.H., PITT, D.G. and EVANS, G., "A lifespan developmental study of landscape assessment", *Environmental Psychology*, 3, 1983, pp. 115-128.

66. HOLCOMB, B., "The perception of natural vs. built environments by young children", in *Children, Nature and the Urban Environment*, Proceedings of Symposium, May 19-23, 1975 at George Washington University, Washington, D.C., pp. 33-53.

67. LYONS, E., *op. cit.*, (reference 22).

68. BALLING, J.D. and FALK, J.H., *op. cit.*, (reference 12).

69. PENNING-ROWSELL, E.C., "A public preference evaluation of landscape quality", *Regional Studies*, 16, 1982, pp. 97-112.

70. BALLING, J.D. and FALK, J.H., *op. cit.*, (reference 12).

71. LYONS, E., *op. cit.*, (reference 22).

72. TUAN, Y-F., *op. cit.*, (reference 17).

73. SHEPARD, P. and MCKINLEY, D., *The Subversive Science: Essays Toward an Ecology of Man*. Boston: Houghton Mifflin, 1969.

74. ROWLES, G.D., *Prisoners of Space? Exploring the Geographical Experience of Older People*. Boulder, Co.: Westview Press, 1978.

75. MACIA, A., "Visual perception of landscape: sex and personality differences", in ELSNER, G.H. and SMARDON, R.C. (eds.), *Our National Landscape*. USDA, Pacific Northwest Forest and Range Experiment Station, Berkeley, California, 1979, pp. 279-285.

76. DEARDEN, P., *op. cit.*, (reference 50).

77. PETERSON, G.L., "Recreational preferences of urban teenagers: the influence of cultural and environmental attributes", in *Children, Nature and the Urban Environment*. Proceedings of Symposium, May 19-23, 1975, George Washington University, Washington, D.C., pp. 113-121.

78. LYONS, E., *op. cit.*, (reference 22).

79. TUAN, Y-F., *op. cit.*, (reference 17), p. 53.

80. KAPLAN, R., "Predictors of envrionmental preference: designers and clients", in PREISER, W. (ed.), *Environmental Design Research*. Stroudsburg: Dowden, Hutchinson and Ross, 1973, pp. 265-274; BUHYOFF, G.J., WELLMAN, J.O., HARVEY, H. and FRASER, R.A., "Landscape architects' interpretation of people's landscape preferences", *Journal of Environmental Management*, 6, 1978, pp. 255-262.

81. For example, DEARDEN, P., *op. cit.*, (reference 50).

82. KAPLAN, S. and KAPLAN, R., *op. cit.*, (reference 14)

83. DANIEL, T.C. and BOSTER, R.S., *op. cit.*

84. For example WALLACE, B.C., "Landscape evaluation and the Essex coast", *Regional Studies*, 8, 1974, pp. 299-305; ROBINSON, D.G., LAURIE, I.C., WAGER, J.F. and TRAILL, A.C., *Landscape Assessment*. Manchester: University of Manchester, 1976; DEARDEN, P., *op. cit.*, (reference 6).

85. DEARDEN, P., "Public participation and scenic quality analysis", *Landscape Planning*, 8, 1981, pp. 3-19.

86. ULRICH, R.S., *op. cit.*, (reference 14), p. 118.

87. However, see research by CRAIK, K.H., "Individual variations in landscape description", in ZUBE, E.H., BRUSH, R.O., and FABOS, J.G., *op. cit.*, (reference 8), pp. 130-150; and MACIA, A., *op. cit.*, (reference 75).

PLATE 9 Cathedral Grove is one of the last remaining stands of old-growth Douglas fir in British Columbia. ►

PART II

APPROACHES

PLATE 10 Vancouver, British Columbia, as seen from the Spanish Banks. ►

4 LANDSCAPES OF EXPERIENCE: MALCOLM LOWRY'S BRITISH COLUMBIA

J. Douglas Porteous

Now there is an empty beach and beside it a park with picnic tables and tarmac access; the sea air stinks with car exhaust. And the city that ignored him plans to cement a bronze plaque in his memory to the brick wall of the new civic craphouse.

Earl Birney (*Selected Poems of Malcolm Lowry*, 1962) on Vancouver's belated tribute to Canada's greatest novelist.

INTRODUCTION

Literary geography has become a firmly-established area of endeavour only during the last decade, despite a history extending back to the beginning of the century.[1] This re-engagement has come about, negatively, as a reaction to the sterility of the New Geography of the 1960s, and positively, as a substantive research area for geographers involved in the "humanistic revival" which began in the 1970s. The objectives of humanistic literary geography are varied, but at the very minimum, "if the reading of literature assists us in seeing the landscapes of the world around us in a clearer light then we have achieved our modest goal."[2] The literary geographer's aim is to understand places, and the relationships between people and places. His data are the data of experience.

Human experience is intimately related to place.[3] Traditionally, literary geographers have emphasized the importance of the regional novel's sensitive evocation of place, and we have been inundated by studies of such placeful novelists as Thomas Hardy, Mary Webb, Arnold Bennett, Francis Brett Young, William Faulkner, D.H. Lawrence, and Willa Cather.[4] "Sense of Place" has become a geographical shibboleth, hardly as yet disturbed by Eyles' behavioural study demonstrating that, for one English town at least, only a minority of interviewees had a sense of place which emphasized "way of life" or "roots."[5]

Despite the evidence amassed by those involved in the recent interest in "valued environments"[6] that attachment to place is vital for well-being, the growth in spatial and social mobility during the present century has resulted in a growing detachment of people from their "roots" in place.[7] For refugees, exiles, transients, immigrants, and even that fifth of the North American population which moves house each year, *declassement*, *deracinement*, and *depaysement* have become normality. The sad content of Johnston's optimistic "intimate journey" belies the title *You Can Go Home Again*;[8] Thomas Woolfe and the Beatles were right — you can't.

Attachment to place, therefore, is only one side of the experiential dialectic. Despite the emphasis of humanistic geographers on such fundamental antinomies as place and space, insider-outsider, native-nonnative, place and placelessness, roots and rootlessness, and home and journey,[9] most literary geographers continue to discuss the regional novelist's fascination with place, insider, native, roots, and home. Very few have confronted the existential problems of the outsider, the nonnative, the rootless, placeless and homeless.

Rootedness in place spells security. Yet for many security must be abandoned, at least temporarily, in the quest for adventure, the loosening of childhood restraints, and the assuaging of curiosity. There is, indeed, a frequently encountered love-hate relationship with home.[10] This lack of balance in literary geography led me to develop an experimental matrix for the study of novels and poetry (Table 1,4), the rationale for which is explained elsewhere.[11]

The majority of geographical work on literary themes falls into the "home-insider" cell. There has been very little concern with negative feelings about home (home-outsider); with the very general modern feelings of angst, anomie, and alienation experienced especially in cities or while in exile (away-outsider); or the rarer effusions of those who feel "at home" only when travelling, such as Gavin Maxwell or Wilfred Thesiger. In order to remedy this situation, I called for a redirection of literary geography towards the three neglected cells and suggested that the novels and travelogues of Malcolm Lowry, Graham Greene, and V.S. Naipul would be a good starting points.

One must practice what one preaches; and for this volume the most appropriate author for analysis is Malcolm Lowry, Canada's greatest novelist. Lowry's work falls almost entirely into the away-outsider mode, with the sea-novel *Ultramarine* (1933) as an example of the away-insider approach. Lowry clearly felt entrapped in his parental home (home-outsider). And, as in much of the literature of "away", including Graham

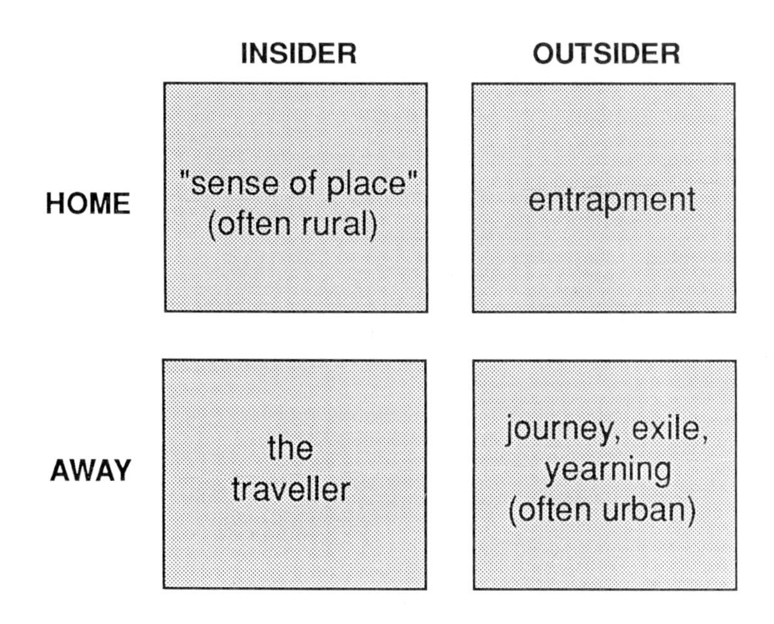

Table 1,4 An Experimental Matrix for Literary Geography

Greene's relentless studies of exiles, there is in Lowry's works a desperate yearning for a new home-place. This Lowry achieved in Dollarton, British Columbia.

Elsewhere I have discussed Lowry's "inscape", a landscape of the mind which led to the equation of volcanic Mexico with hell, and wilderness British Columbia with heaven.[12] This essay will deal with three themes in Lowry's work which have relevance not only to the theoretical discussion above, but also to the environment of British Columbia. These themes are: sea and land, the major physiographic division of the earth's surface, which relates to Lowry's brief away-insider mode and his eventual reconciliation with littoral life; home and journey, a fundamental psychic distinction incorporating Lowry's away-outsider mode, which develops into home-insiderhood only in British Columbia; and city and wilderness, where Lowry rejects the alienation of cities, and in particular Vancouver, British Columbia, in favour of the Canadian coastal forest. First, however, there must be a brief contextual discussion of Lowry's life and work.

THE LIFE AND WORK OF MALCOLM LOWRY

> A day of sunlight and swallows ...
> And saw the fireman, by the fiddley, wave.
> And laughed. And went on digging my own grave ...
>
> M. Lowry, *Hostage*

Alcoholic and manic-depressive, infantile, orally-fixated, sexually insecure, narcissistic, generally inept, and ultimately a suicide, Malcolm Lowry (1909-57) was also a literary genius.[13] Although he published only two novels during his lifetime, the second, *Under the Volcano*, is a masterpiece. His whole opus (Table 2,4) has much to say to students of the landscapes of experience.

Malcolm Lowry's childhood was spent in a wealthy suburban setting on the Cheshire side of the Liverpool conurbation in England. His father was a successful businessman, a stern, low-church teetotaler, a very threatening superego; Lowry went in fear of authority all his life. Lowry had an ambivalent feeling for his mother, who at one point rejected him and who often left with her husband on business trips. Sent to boarding school at age seven, he remained thus institutionalized until escaping to sea between school and university. Lowry was a loner; in Mexico he prayed to the Virgen de la Soledad, "the Virgin for those who had nobody them with." He was a heavy drinker from his teens, a spinner of tales from school magazine onwards, an anxious neurotic since his early alienation from the deeply Calvinistic austerity of his family home. All his life he rejected authority, but craved dependence, a difficult psychic balancing act hardly conducive to serenity.

Lowry's undergraduate novel, *Ultramarine* (1933), was based on his recent experiences as a deckhand en route to the Far East. During his time at Cambridge he visited novelist Nordahl Greig in Norway and conceived the novel *In Ballast to the White Sea* which was unfortunately destroyed by fire while still in manuscript. After Cambridge he visited France and Spain, married, and travelled to New York. There, already an alcoholic, he was admitted to the psychiatric wing of the Bellevue Hospital, an experience which later figured in the novella *Lunar Caustic* (1968). A failure in Hollywood, he removed to Mexico where his eighteen-month stay in Cuernavaca provided material for his masterwork, *Under the Volcano* (1947). Abandoned by his wife, thrown out of Mexico for drunken excesses, he met his second wife in Los Angeles and retired with her to a squatter's shack at Dollarton, on the north shore of Burrard Inlet opposite the city of Vancouver. Here they lived from 1940 to 1954, and Dollarton figures, as Eridanus, in all the post-*Ultramarine* novels. Dollarton became

Table 2,4 THE PUBLISHED WORKS OF MALCOLM LOWRY

	Title	Date of Publication	Date of Experience	Lowry's Age	Chief Settings
1	*Ultramarine*	1933	October 1927-September 1928	18-19	Far East (shipboard)
2	*Lunar Caustic*	1968*	June 1935	26	New York (Psychiatric ward)
3	*Under the Volcano*	1947	November 1936-July 1938	27-29	Mexico; B.C. references
4	*Dark as the Grave Wherein My Friend is Laid*	1969*	November 1945-May 1946	37	Mexico; some B.C.
5	*Hear Us O Lord From Heaven Thy Dwelling Place*	1961*	November 1947-January 1949	38-40	British Columbia, Panama, Europe
6	*October Ferry to Gabriola*	1970*	1945-46; October 1946	36-38	Ontario, British Columbia

* Posthumous

the home-base for a succession of trips to Ontario (1944), Mexico (1945-46), from which he was again expelled, Haiti, New York, Ontario again (1946-47), France and Italy (1947-49). All these journeys figured in the four novel-travelogues entitled *Hear Us O Lord From Heaven Thy Dwelling Place* (1961), *Dark As The Grave Wherein My Friend Is Laid* (1969), *October Ferry To Gabriola* (1970), and *La Mordida* (unfinished, unpublished).

In 1954 Lowry was forced to leave Dollarton, travelled to Italy in an unsuccessful search for an alternative home-place, underwent psychiatric treatment in London, and settled in 1956 in the village of Ripe in Sussex. His suicide there was recorded as "death by misadventure," the latter word being an inadvertently accurate judgement on the whole of Lowry's life. The reader of the six published novels is immediately struck by certain themes which recur throughout. Lowry's novels are autobiographical to the highest degree. His only material was his own experiences, his own feelings, his own journeyings, his own anguish. In every work the protagonist is Lowry himself, with only the slightest changes of detail, indeed, in at least one short story the author slips from "Sigbjorn Wilderness," the protagonist, to "I" and back again. Many of Lowry's characters bear Norwegian or Celtic names; to Lowry these peripheral nations were breeders of strong, heroic, troubled seafarers, a role he took on early in life.

The theme of the sea, unfettered, open, clean, but often violent, runs through most of his works. On land, if the seashore is not available, volcanoes are preferred. Volcanoes and storms at sea fascinated Lowry. His own character involved a good deal of suppressed violence, and he displayed bouts of high energy and indiscipline followed by passive quiescence, a characteristic he shared with both volcanoes and the sea. His identification with landscape features is not unexpected; a psychiatric examination suggested that Lowry was unable to distinguish fully between himself and his environment.

Further, all the novels are journey novels. These may be actual, physical journeys across landscape or seascape, or they may be interior journeys through the landscapes of the mind. Frequently, they may be read at both levels. Within the journey theme, the concepts of home and homelessness recur. The traveller is frequently unhappy with his journeying, he seeks a place which he may call home. Once found, however, home can be threatened by outside forces and the new-settled traveller fears its loss. Throughout this repeated dialectic of journey and home Lowry and his protagonists appear as existential outsiders, achieving a brief insiderhood only in the squatter settlement at Dollarton.

With respect to his life in Dollarton, perhaps it is worthwhile here to emphasize the Canadian-ness of Malcolm Lowry. It has been noted that "the most important novel ever written in Canada is a novel about Mexico (*Under the Volcano*) written by an Englishman in exile" and that Canada was Lowry's "last stand."[14] Mexican *Volcano* is ultimately pessimistic, and a reading of this novel alone has frequently coloured Canadian literary views of Lowry. But when we read his poems about Burrard Inlet, and his Canadian novellas and stories we realize that Lowry:

> ... is not in fact writing about Canada as a transient outsider. He is writing about it as a man who over fifteen years lived himself into the environment that centred upon his fragile home where the Pacific tides lapped and sucked under the floorboards, and who identified himself with that environment...[15]

Perhaps some of the Canadian antipathy towards immigrant Lowry is explained by the unflattering fact that: "It was really landscape more than culture that influenced him in Canada."[16] In this sense Lowry is a geographer's novelist in general, and a Canadian geographer's novelist in particular.

SEA AND LAND

> Resurgent sorrow is a sea in the cave
> Of the mind —
> Abandon it! ...take a trip to the upper shore. Lave
> Yourself in sand; gather poppies; brave
> The fringe of things, denying that inner chasm ...
>
> M. Lowry, *Whirlpool*

Sea:land is the ultimate antinomy in physical geography. The two are complementary; they interpenetrate. Yet human beings are likely to be placed on the one, and with few expectations become placeless voyagers upon the other.

Lowry's fascination with the sea had deep roots. A ship-owning grandfather was fantasized as a swash-buckling sea captain. The author's father was a bemedalled lifesaver. Lowry, himself an excellent swimmer, grew up close to the Irish Sea and to the great shipping lanes which focussed upon Liverpool. His youthful reading included the wandering seafarer tales of Conrad, Melville, O'Neill, and Jack London. His more immediate literary mentors were Nordahl Greig and Conrad Aiken, both known for their sea novels. Almost all Lowry's novels contain a seaborne theme. *Ultramarine* is a young-man-finds-himself-at-sea novel fairly typical of the age. The protagonist Dana (taken from Dana's *Two Years Before the Mast*) is a "toff" who goes to sea to confront existential extremes and, through suffering them, to find himself. His ship, significantly, is named the *Oedipus Tyrranus*. From the landlubber's viewpoint, the unrestricted placelessness of the sea, and the irresponsible adventures in short-stay ports, seem enviably romantic. But the tyro quickly discovers that life at sea is, ultimately, boring. The work is hard and the deckhand is too tired to contemplate the beauties of sea and sky. *Ultramarine*[17] lacks purple passages. Moreover, port life is disappointing — "how alike all these harbours are" (p. 145) — and consists largely of drunken binges, pointless conversations, and encounters with raddled whores. Ultimately, sailors are happiest when they are going home, yet on land they are homeless and must compensate in alcoholic excess: drunk is the sailor, home from the sea.

Nevertheless, we can grasp Lowry's early positive feelings for the sea, specially in his evocation of ocean life while in some squalid Far Eastern port. "O God, O God, if sea life were only always like that! If it were only the open sea, and the wind racing through the blood, the sea, and the stars forever!" (p. 77). It is the sea wind that stirs Lowry, "The wind! The wind! The cold clean scourge of the ocean" (p. 66). And, on reaching home, there is the ever-present urge to set sail once more (p. 185):

> To sail into an unknown spring, or receive one's baptism on storms' promontory, where the solitary albatross heels over in the gale, and at last to come to land. To know the earth under one's foot and go, in wild delight, ways where there is water ... to return again over the ocean ... at last again to be outward bound, always outward, always onward ...

This evocation of the spell of the sea must stir even the most inveterate landlubber.

Yet Lowry's most vivid evocations in *Ultramarine* are not of the sea itself, but of the filthy, odorous reality of shipboard life. Throughout the voyage Dana's gaze is not outward at the sea, or upward at the sky, but downward into the hold, the crew's quarters, the O'Neill-like inferno of the engine room (p. 158), which is a:

> Maelstrom of noise, of tangled motion, of shining steel ... where the red and gold of the furnaces mottled the reeking deck, and the tremulous roar of the cage's fires dominated a sibiliant, continual sputter of steam. The *Oedipus Tyrranus's* firemen ... half naked, gritty and black with coal, and pasty with ashes, came and went in blazing light, and in the gloom; flaming nightmares, firelit demons. The furnace doors opened, and scorpions leapt out, spirals of gas spun and reeled ...

In this frightening vision Lowry has caught his first glimpse of the abyss, the inferno, into which he and the chief character, in *Volcano*, finally fall.

This single voyage proved a turning point in Lowry's life. From thence he perfected the role of the rough, drunken sailor among his university and literary acquaintances. With the publication of *Ultramarine* he could pose as the doomed sailor of genius. The swaggering Norwegian persona he adopted at this time (a kind of Joycean "Scandiknavery") led Lowry to lie under tables with a bottle to recreate the feelings of life at sea.

All the remaining books have explicit or submerged sea themes. They are replete with the resounding names of ships, the exotic names of ports. In *Under the Volcano*[18] both the Consul, the protagonist, and his brother (both aspects of Lowry) have been seafarers. The seaborne ambience, clean, sleek, silent, of their memories and wishes clashes sharply with the arid, maddening hell of interior Mexico, where the sea can appear only in metaphor, as when the Consul hears "a soft wind-blown surge of music, from which skimmed a spray of gabbling, that seemed not so much to break against as to be thumping the walls and towers of the outskirts; then with a moan it would be sucked into the distance" (p.17).

Again, in *Lunar Caustic*[19] the inmates peer through the windows of New York's misnamed Bellevue asylum at the freedom of a waterway with ferry and motorboats. Yet they identify chiefly with a wrecked coal barge, "sunken, abandoned, open, hull cracked, bollards adrift, tiller smashed," going nowhere. The chief character, Bill Plantagenet, sees himself as a ship. Later, writing in despair from Mexico, Lowry described himself as a sinking ship.

The titles *In Ballast* and *October Ferry To Gabriola* are further indication of this love affair with the sea. A chapter in the latter, "The Tides of Eridanus," lovingly evokes Lowry's feelings for the calm:menacing, flowing: ebbing, eternal:transient, silent:clamorous sea on or by which he spent the greater part of his life. Doomed to live somewhere on land, he chooses a war-time squatter shack at Eridanus (his name for Dollarton) where, though:

> whole cities, countries be wiped out, ... Eridanus, with its eternal fishermen and net-festooned cabins bordering the inlet ... whose ceaseless wandering yet ordered motions were like eternity, looked ... somehow transported straight to heaven. Eridanus *was*. (p. 79)

Hear Us O Lord From Heaven Thy Dwelling Place[21] is made up of seven short stories or "meditations." The first, "The Bravest Boat," begins with a recurrent Lowry image, "a day of spindrift and blowing sea foam" (p. 11). In hostile Vancouver, Sigurd Storlesen seeks comfort by Stanley Park's

Lost Lagoon. Renamed Sigbjorn Wilderness, he leaves for Europe on a freighter. Throughout, Wilderness considers himself as the Ancient Mariner, complete with albatross. The passage, "Through the Panama" is replete with the sexual imagery of land-locked water. Wilderness's dream of the *vagina dentata* is matched by the passage through the canal, whose locks snap shut around the ship, and whose banks bear a dense, fetid tropical forest. Reborn, Wilderness finds a severe Atlantic storm exhilarating in comparison with the canal passage, and is able to come to some form of self-realization. After an unsuccessful European tour, the writer and his wife return to Eridanus, girded by mountains, fronted by the sea, a tidal, boat-bedecked, full-encrusted, belled and fog-horned, surfed and drift-wood-laden Dylan Thomas-ish nautical place inhabited year-round only by beached sailors, Celts and Scandinavians all. The novel ends in rain:

> And the rain itself was water from the sea, as my wife first taught me, raised to heaven by the sun, transformed into clouds, and falling again into the sea. While within the inlet itself the tides and currents in that sea returned, became remote, and becoming remote, like that which is called the Tao, returned again as we ourselves had done ... Three rainbows went up as rockets across the bay: one for the cat ... laughing we stooped down and drank.(p. 286)

That the sea may be the transforming agent in the life of a protagonist is a literary cliche. But in Lowry the sea is more than merely symbolic. It has an omnipresent, magical quality. Lowryan characters sail over it, live beside it, listen to it, dream of it. They need the sea. In Lowry the sea is always positive. It challenges, but never harms. It aids protagonists in their search for self-realization. In Lowry's words, the sea is "on the side of life."

In contrast, land is, at best, ambivalent. Gardens are cool and inviting, or like the Borda Gardens of Cuernavaca, may be corrupt and menacing. Forests are sheltering; they surround and nurture the cabin at Eridanus. But they may also be menacing, harbouring wild beasts and tangled roots which actually seem to conspire in the downfall of the traveller. Roads lead to exhilarating or devastating experiences. In particular, paths in forests are used self-consciously by Lowry as a symbol of the unfathomable ambivalence of life on land. In one of his most lyric stories, "The Forest Path to the Spring," in *Hear Us O Lord* (p. 272), we find that:

> There has always been something preternatural about paths, and especially in forests ... for not only folklore but poetry abounds in symbolic stories about them: paths that divide and become two paths, paths that lead to the golden kingdom, paths that lead to death ...

But best of all, forests lay between Eridanus and the city, and it was the forest and its paths, impenetrable by automobile, which kept civilization at bay.

Like the forest, the mountain rim north of Eridanus was protective; these mountains could be captured in their reflection in the sea. But again, the fascination with volcanos which took Lowry to Popacatepetl, Ixtaccihuatl, Vesuvius, Etna, and Stromboli is a fascination with the inferno. Shortly before his death, Lowry spent a week on Stromboli, awed by the black sand, and the half-ruined villages, and the regular eruptions. In *Under the Volcano* the brooding presence of Popacatepetl cannot be escaped, and it is below this volcano that the Consul's life draws to its drunken close in a symbolic mountain cavern, the ironically-named cantina El Farolito (The Lighthouse).

Volcano is Lowry's land novel. The landscape of Cuernavaca is richly symbolic, a "forest of symbols" in the sense of Baudelaire. The idea of water is there, in memories, in wishes, in Cuernavaca's myriad swimming pools. But little rain falls; fountains and even bathroom showers run dry. If the sea to Lowry is on the side of life, in Cuernavaca it is far off; the liquids encountered by the Consul are invariably alcoholic. The Borda Gardens, luxuriant retreat of Maximilian and Carlotta, are stagnant, ruined, obscenely rotting. A forest girds Cuernavaca, but far from providing shade and coolness, it is the scene of darkness, ruin, and confusion. Through it run paths that lead only to perdition, a far cry from the idyllic British Columbian "Forest Path to the Spring."

Lowry was clearly no happier with life in an inland situation than with life on the ocean wave. The ideal compromise was a littoral way of life. In British Columbia, however, he still had to make the decision between living in a coastal city or in the nearby wilderness seashore. City versus wilderness, a common theme in North American history, became a major issue in Lowry's life and work. Here the chief contrast lies between Vancouver (Lowry's Enochvilleport) and the squatter settlement at Dollarton (Eridanus).

CITY AND WILDERNESS

> Beneath the Malebolge lies Hastings Street
> The province of the pimp upon his beat,
> Where each in his little world of drugs or crime
> Moves helplessly or, hopeful, begs a dime
> Wherewith to purchase half a pint of piss —
> Although he will be cheated, even in this ...
>
> M. Lowry, "Christ Walks in the
> Infernal District too"

Uneasy in continental interiors, far from the presence of the sea, Lowry was also uneasy in built environments. On a bus trip across the United States he recorded only "slag heaps, automobile accidents, billboards, men's rooms in bus stations and the graffiti therein, and suchlike instances of civilized nastiness."[22] The Mexican dystopia is seen as a plethora of signs, posters, vultures, scorpions, hideous pariah dogs, ragged and filthy natives, twisting roads and railway lines, doubtful cantinas, and above all, Cuernavaca's Ferris Wheel of life and death, on which the Consul is suspended upside down. These urban and Mexican dystopias reflect the dark side of Lowry's inner landscape.

Cities, for Lowry, were at best placeless. Only rarely does Lowry record his feelings for European or Eastern cities; in Mexico, where descriptions of urban scenes are common, Lowry notes that "the names were the only beautiful things, for the towns were all the same."[23] Lowry, of course, was chiefly familiar with seaports. His feelings of their rather squalid, seedy sameness, an atmosphere also common in the contemporary novels of Graham Greene, was carried over to that modern equivalent of the seaport, the appropriately-named airport. As early as 1945 Lowry was expressing the common modern perception that all airports look alike. In Vancouver, Los Angeles, Mexico, the airport provided the same peculiar view of the attendant city, "a few distant factory chimneys, a Gogolian landscape."[24]

Yet on entering a city, Lowry became immediately subject to another modern, but usually undiagnosed problem, that of urban sensory overload.[25] Lowry was extremely sensitive to city noise and fearful of crowds and traffic. Rome was "an inferno with twelve different kinds of buses coming at you from every direction;"[26] Mexico City "seemed much the same, smells, noise ... with which went the same invitation to get out of it as soon as possible."[27] Villahermosa fails to live up to its name, while retreat to the north only brings Lowry into contact with the "rending tumult of American cities, the noise of the unbandaging of great giants in agony."[28]

One may readily adapt to aesthetic ugliness, placelessness and, eventually, to sensory overload. Yet, for Lowry, the city is the scene of a far worse phenomenon; the subjection by human beings of nature.

Cities are inorganic, distant from nature.[29] Consider just one sensory modality. The smell of Mexico City, of "gasoline, excrement, and oranges"[30] contrasts favourably with the "disgusting smell" of Vancouver city buses.[31] Both are vile in comparison with "the smell of salt, pines, and evening smoke"[32] in Dollarton or, in the evening outside the squatter shack, "the rich damp earth, myrtle and the first wild crabapple and wild cherry blossoms, all the wild scents of spring, mingled with the smell of the sea and from the beach the salt smells, and the rasping iodine smell of seaweed."[33]

Lowry celebrates wild nature from his retreat only fifteen miles from Vancouver, where he and his wife live in a hand-built shack surrounded by forest and sea, row their boat, fetch water from the spring, lie awake on winter nights in fear of high tides and storms, meet cougars and less fearsome forest denizens face to face, and like the Ancient Mariner, come to "reverence all things that God made and loveth."[34] Here one can build one's house from materials salvaged from sea or forest, a house which seems an organic part of the forest. This is the simple life of which the Consul, trapped in Cuernavaca, and doomed to penetrate *Under the Volcano*, dreams. The whole of the novella "The Forest Path to the Spring" is a paean of praise to this lifestyle. At Eridanus, Lowry had become an insider in an almost changeless, timeless, existential world, "a region where such words as spring, water, houses, trees, vines, laurels, mountains, wolves, bay, roses, beach, islands, forest, tides and deer and snow and fire, had realized their true being, or had their source ..."[35]

From Eridanus Vancouver could not be seen, but an oil refinery complex was visible, and grew over the years. At first viewed only in aesthetic terms, the refinery eventually becomes a threat to Eridanus. With "the assault of unique oil-smells" the growth of oil-slicks, and its constant flare-off, Shellco becomes a "livid flickering City of Dis."[36] Above the refinery, a coarse cerise neon sign, one letter burnt out, reads "HELL".

Beyond Shellco lay Vancouver, of which Lowry has only one positive statement to make, regarding the ready availability of books from the city library.[37] Otherwise, the inorganic, disorganized nature of cities was, for Lowry, typified by the city of Vancouver. In the first story in *Hear Us O Lord*, Lowry's protagonist, about to escape by sea, provides a devastating comment on what he is escaping from, a city symbolically renamed Enochvilleport, after the son of Cain, and:

> composed of dilapidated half-sky-scrapers, at different levels, some with all kinds of scrap iron, even broken airplanes, on their roofs, others being moldy stock exchange buildings, new beer parlours crawling with verminous light even in mid-afternoon and resembling gigantic emerald-lit public lavatories for both sexes ... totem pole factories, drapers' shops with the best Scotch tweed and opium dens in the basement ... cerise conflagrations of cinemas, modern apartment buildings, and other soulless behemoths ... [38]

This is a truly apocalyptic vision, the destruction by human agency of the beauty of "a harbour more spectacular than Rio de Janeiro and San

Francisco put together."[39] Deep within this urban jungle, Lowry sees "occasional lovely dark ivy-clad old houses that seemed weeping, cut off from all light, on their knees ..."[40] The organic, the human scale, is already falling prey to the monolithic sameness perpetrated by naive imitators of Le Corbusier and Mies van der Rohe.

Adaptation to the wilderness did not come easy to Lowry, but adaptation to urban life was impossible. A brief stay in a Vancouver apartment house is recounted in *October Ferry*. Here Lowry witnessed the destruction of the old West End to make what is now one of the highest-density clusters of apartment towers in the western world. A self-righteous cleanup campaign relieves inner Vancouver of trees and "poor old steamboat Gothic buildings" (p. 177) which, 30 years later, a more enlightened city government might be struggling to preserve and rehabilitate. "Poor old Vancouver!," thinks Ethan Llewelyn, as bulldozers uproot trees and house wreckers torture a house to pieces "bit by bit ... almost as if they meant to eat it," with a "calamitous wrenching sound" (p. 187). The replacements are "fake modern buildings ... soulless Behemoths in the shape of hideous new apartment buildings, yet more deathscapes of the future" (p. 177). The only pleasant views in Vancouver are the vistas, between buildings, of the surrounding mountains.

Yet even these do not escape the machinations of the city. In Lowry's view, Vancouver is to be condemned not merely for its scabrous self, nor yet for its hideous treatment of heritage buildings, but rather for its usage of the surrounding landscape:

> anyone who had ever really been in hell must have given Enochvilleport a nod of recognition, further affirmed by the spectacle ... of the numerous sawmills, relentlessly smoking and champing away like demons, Molochs fed by whole mountainsides of forests that never grew again, or by trees that made way for grinning regiments of villas in the background of "our fair, expanding city," mills that shook the very air with their tumult, filling the windy air with their sound as of a wailing and gnashing of teeth; all these curious achievements of man, together creating ... "the jewel of the Pacific" ...

To live in cities was, for Lowry, an abomination. In the forest, beside the sea, with "the wonderful cold clean fresh salt smell of the dawn air, and then the pure gold blare of light from behind the mountain pines, and the two morning herons ..." one was comparatively safe. And in such a spot there was always the chance of "waking to the sudden O'Neillian blast of a ship's siren, taking your soul to Palembang in spite of yourself ..."[42]

Ultimately, for Lowry, the city is a predator, a creator of "deathscapes." All three novels based in Eridanus, *Hear Us O Lord*, *Dark As The Grave*, and *October Ferry*, contain this disturbing word. Deathscapes result when urban expansion destroys non-urban landscapes; in Vancouver's case, city expansion is capable of destroying wilderness as well as rural environments.

Urban fringe development can create topophobia, a landscape of fear.[43] Fear of the city and its depredations grows throughout the later Lowryan novel sequence. In *Dark As The Grave* there is a single experience of it, the fear that when the Lowrys return from Mexico they would find their squatter shack on government land had been destroyed in favour of "auto camps of the better class."[44] The fear continues in *Hear Us O Lord*, where a newspaper report of an industrial boom in British Columbia is juxtaposed with a report of the sudden withering of a thousand-acre forest; "Foresters are unable to explain it, but the Indians say the trees died of fear but they are not in agreement about what caused the fright."[45] The fear of civilization is inferred. Fear becomes paramount in *October Ferry* when developers and government proceed to evict the Eridanus squatters in favour of subdivisions and industry. The Llewelyns, now exiles, seek a new homeplace, but are bewildered by ugly hoardings, the unpleasant city of Nanaimo, (Victoria, a rather beautiful city, is significantly not mentioned), crass real estate advertisements, and modern "ugly, standardized houses ... crowded together in an exact row as though dropped there by a conveyor belt."[46]

City people hate, fear and envy the Dollarton squatters. The squatter settlement and its forest environs is wasted on the squatters, who do not pay taxes, who lower the dignity of the city, who live in a rat's nest of vice and crime, whose land should be a public park, and whose shacks, "like malignant sea-growths should be put to the torch." "Squatters! The government's been trying to get rid of them for years ... cut down all those trees, open it up to the public, put it on the map."[47] Already parts of Eridanus had become "a suburban dementia" behind which "was still the dark forest, waiting, one hoped, for revenge."[48] Rosslyn Park Real Estate and Development Company ("Enquire Here, Scenic View Lots") attracts urbanites in cars, bringing "an all but continuous uproar," drives away wildlife, and produces a "big new schoolhouse, a great concrete block of mnemonic anguish," amidst "a hideous slash of felled trees, bare broken ugly land crossed by dusty roads and dotted with new ugly houses where only a few years ago rested the beautiful forest that they had loved."[49]

From the city as predator there is no escape, for Lowry realises at last that it is civilization that is the enemy:

> Canada was indeed a pretty large country to despoil. But her legends, nearly all her most valuable and heroic history

> of spoilation ... the conquering of wilderness ... was part of [man's] own process of self-determination ... progress was the enemy, it was not making man more happy and secure. Ruination and vulgarization had become a habit.[50]

In the face of urban philistinism and greed Lowry capitulates, anticipating Lynn White in his vision of the future as "a picture of Progress in the form of Jesus Christ driving a locomotive across a virgin forest."[51]

Civilization is the creator of deathscapes.[52] Ultimately, Lowry indicts machine civilization for its greedy, remorseless devastation of the natural, the organic. Unfortunately for Canada, this indictment, although meant to be general, is expressed by Lowryan protagonists wholly in Canadian terms. Canada, to Lowry in the 1940s, could still be redeemed; unlike Europe, or even the United States, its wilderness landscapes still remained relatively untouched.

Some of the criticism sets the scene for an understanding of Canadian philistinism. Typical of Lowry, the milder deathscape scenes are alcoholic in tone. Sexual segregation in drinking establishments is simply one of the "fatuous prohibitions" of this dour, rather Scottish nation. Canadian beer parlours, in their "ghastly malevolence ... summed up in that one genteel funereal substantive" are "the largest and ugliest drinking establishments on earth,"[53] further expression of a dull, dour, prevailing presbyterianism. Vancouver, source of ultimate misery to Lowry, can from the distance of Mexico be seen almost light-heartedly as having:

> a sort of Pango Pango quality mingled with sausage and mash and generally a rather Puritan atmosphere. Everyone fast asleep and when you prick them a Union Jack flows out. But no one in a certain sense lives there. They merely as it were pass through. Mine the country and quit. Blast the land to pieces, knock down the trees and send them rolling down Burrard Inlet ...[54]

But once engaged in the Vancouver scene, Lowry was compelled to acknowledge both the despair, degradation and meanness he perceived in everyday urban life and the desperate dilemma in which British Columbians, with a natural wilderness on their doorstep, find themselves trapped:

> Vancouver, Canada, where man, having turned his back on nature, and having no heritage of beauty else, and no faith in a civilization where God has become an American washing machine, or a car he refuses even to drive properly — and not possessing the American elan which arises from a faith in the very art of taming nature herself, because America

> having run out of a supply of nature to tame is turning on Canada, so that Canada feels herself at bay, while a Canadian might be described as a conservationist divided against himself ... [55]

As he later observes, British Columbia is "not styled B.C. for nothing."[56] The heart of the matter is that postwar Canadians had become arrant modernizers, creators of placeless deathscapes, destroyers of wilderness in pursuit of profit. And the cruel truth was that "Canada's beauty was in that wilderness ... It was the only originality it had."[57]

Canadian despoliation, of course, is largely a reflection of American drive and profiteering. Canadian deathscapes merely echo the "barren deathscapes of Los Angeles."[58] The end result is large-scale urbanization and the ultimate and ironic divorce of Canadians from their natural surroundings. Although never mentioning television, Lowry had a clear vision of the electronic future, a world where:

> It was as if they had exchanged sunlight on water for photographs of sunlight on water, cool commotion of blowing grasses and pennyroyal, or reeds and rippling waters [or] migrating birds, for the tragic incidental music that always accompanies documentaries involving blowing grasses, rippling waters and migrating birds, and soon they would not be able to have told the difference ... [59]

Lowry's metaphor for this *National Geographic* of the soul is revelatory, archetypal, and apocalyptic; deathscape is the city in the wilderness, "the abomination of desolation sitting in the holy place."[60]

HOME AND JOURNEY

The above discussions of the sea:land and city:wilderness antinomies in Lowry's work deal with the major facets of physical and human geography which influenced Lowry's life. On a more conceptual level, and especially in connection with Lowry's seagoing and city-visiting propensities, the overarching theme of home:journey is a useful means of enhancing our understanding of the Lowryan landscape.

Journey

> How did all this begin and why am I here
> at this arc of bar with its cracked brown paint,

papegaai, mezcal, hennessey, cerveza,
two slimed spittoons, no company but fear:
fear of light, of the spring, of the complaint
of birds and buses — flying to far places ...

M. Lowry, "No Company But Fear"

For much of his life Malcolm Lowry was of no fixed abode; in his own words he resided at Hotel Nada. Even his 14 year residence in Dollarton was broken by long journeys. Significantly, and tragically, he called his whole opus "The Voyage that Never Ends." For Lowry was a lifetime fugitive, running from his childhood home, from Europe, from Mexico, finally from Dollarton, and always from himself. That life is a journey is a well-worn cliche; Lowry's life was journey both in reality and metaphor, across real landscapes and within his own mind.

Lowry's schoolboy stories were often set on trains; his jazz preferences include "Going Places" and "Doing Things."[61] Like other middle-class youths, he was allowed a *wanderjahr* in Europe, and the stories he wrote at that time reflect a taste for the exotic. His early life in Paris and London was one of seedy, cheerless apartments, littered with bottles, and with no pretence of permanence. Indeed, in both London and New York he was often so disoriented that he was unable to remember where he lived. The walks he took were pubcrawls. Paranoid, he often felt trailed by unknown followers. He frequently indulged in long wandering monologues which always failed to reach a goal. He speaks, Joyce-like, of his "tooloose-Lowrytrek."

Lowry's life of wandering is an extreme case of the attack of wanderlust experienced by literary British youth between the wars. Compared with more pragmatic Americans, "Britons, confined to a small island, romantically cherish the act and art of journeying for its own sake."[62] Fussell suggests that the outburst of travel fever after World War I was less a case of curiosity than of a blind urge to flee England. D.H. Lawrence and, later, Graham Greene, may be cases in point.

Lowry fled England deliberately, but with few plans. Typically, his central and most formative journeys, to Mexico (1936) and Vancouver (1939) were unplanned. Almost penniless in Los Angeles, Lowry felt Mexico would be cheaper. Flung out of Mexico, and misinformed that a United States visa could only be renewed outside the country, he went to Vancouver. In contrast, the less important early voyages were deliberate adventures, and the later expeditions from Dollarton were either searches for a new home or sad attempts to relive the past. Even Lowry's manuscripts were wanderers; they passed from publisher to publisher by journeys as complex as those described within them.

By middle age this peripatetic Briton had visited the Far East, much of Western Europe and North America, Mexico, and the Caribbean. Wherever possible he went by sea. His initial description of Port au Prince, Haiti, from the sea is a typical traveller's vision — everything seen as an amalgam of somewhere else:

> ... strangely beautiful houses of pointed roofs and of seemingly Norwegian design, church spires here and there rise vaguely in the sun giving it a look of Tewkesbury, while to the right mist lay in pockets of rolling mysterious mountains like Oaxaca ... [63]

Like all travellers, too, he learns that return journeying is dangerous. Places, whether loved or hated, have always changed by the time the traveller, himself also changed, returns.

"The Voyage that Never Ends" was Lowry's framework for a great novel-sequence which was to depict "The Ordeal of Sigbjorn Wilderness." Most of the novels depict journeys, usually voyages. Only *Lunar Caustic* and *Under the Volcano* are physically anchored in a single place. Yet Lowry was nothing if not an introspective novelist. All the novels, without exception, depict psychic journeys, interior journeys from a state of chaos to one of stability in life or death. The interior journey is expressed symbolically as physical journeying, whether actual, as in most of the novels (even *Volcano*, where most of the wandering takes place in one locale), or in terms of journeys of memory or wish, as in *Volcano* and *Lunar Caustic*.

It is the psychic journeying of Lowry which is the more important. Lowry saw himself as Ahasuerus, the Wandering Jew, as Ulysses, as Ahab, as the Ancient Mariner, or as one of the many lonely, doomed wanderers of literature. His voyage into the public psychiatric ward depicted in *Lunar Caustic* (and which he later claimed to be a deliberate "pilgrimage") was a psychic journey meant to cleanse his soul and provide regeneration, explicitly on the lines of Rimbaud's *Saison en Enfer*.

In the psychiatric ward, drying out, he meets his persona face-to-face (the inmate Kalowsky = the Wandering Jew) and experiences the omnipresent existential reality of confinement. It is Bellevue Hospital, after all; the views from its windows are all-important contact with the outside world. Within, the summer heat oppresses; across the river lie the Ice Palace and the Jack Frost Sugar Works. Ships pass by. If coming into port, the inmates let out a cry of hope, partly a shriek, "partly a cheer;" if heading out to sea, the response is dull silence, "as if all hope were heading out with the tide."[64] Throughout, the rotting hulks of coal barges reflect the beached, stranded hopeless madmen who gaze blankly down upon them.

In Ballast, *Hear Us O Lord*, *Dark As The Grave*, and *October Ferry* are explicitly journey novels, where the journey mirrors the mind's wanderings.

Yet *Volcano*, on the surface Lowry's land novel par excellence, is in fact the story of a psychic journey that ends in madness, death, and damnation. It is first of all, a novel of exiles, drawn together in a place which is not merely unsympathetic, but actively hostile. The Consul's inner voice tries to soothe him by telling him he is "only lost, only homeless," as if this were not one of the most terrifying existential conditions. And in *Volcano* the landscape is dynamic; volcanos brood, *barrancas* (ravines) snake through the countryside, forests participate in human destruction. The whole novel is dynamic; it is a wheel, in constant motion, symbolized by Cuernavaca's carnival Ferris Wheel in which the Consul is racked and humiliated. In Lowry nothing is static; motion is of the essence.

"Life is a journey, a passage with no return ... the pilgrim is the man who ... becomes in reality the traveller that everyone is symbolically."[65] Lowry saw his own journey, as reflected in his novels, as Dantesque, with periods of Purgatorio, Paradiso, and Inferno. Paradiso is clearly Eridanus (Dollarton), "the simple life" in British Columbia which the consul can dream of but never attain, and an Eden from which Lowry was finally expelled. Purgatorio is the psychiatric hospital of *Lunar Caustic*. Above all, it is a place of no motion and no hope. For Lowry, the life of the Wandering Jew, though it may lead eventually to hell, is better than the stasis of Purgatory.

So we are left with Inferno, the abyss which Lowry first glimpsed on board ship in *Ultramarine*. The abyss is everywhere, even at sea, for Lowry frequently uses the Maelstrom as metaphor. It is in Oaxaca, the City of Dreadful Night, and in Cuernavaca it is in the rotting gardens, in the fetid garbage-laden *barrancas*, in the deep, dark caverns of its *cantinas*. A majestic setting is of no avail; the *barrancas* lie at the foot of the volcano.

Cuernavaca's landscape is a taut amalgam of psychic journey symbols. Popacateptl is one of those magic mountains which to ascend brings grace. Mountains are postive symbols for the Consul and his wife, whose respective childhoods were spent in Kashmir and Hawaii; the simple life among the mountains of British Columbia is their dream. But the Consul never climbs the sacred mountain. Instead, he is irresistibly attracted to the caverns that lie under the volcano, the dark *cantinas* and the deep *barrancas* that crack apart the town of Cuernavaca. Throughout the novel the *cantinas* and the *barrancas*, one of the latter at the bottom of Lowry's garden in Calle Humboldt, wait, almost sentient, for their prey.

In the last scenes, all the characters lose their way in the Dantesque forest surrounding Cuernavaca:

> Nel mezzo del cammin di nostra vita
> me ritrovai per una selva oscura
> che la diritta via erra smaritta ...[66]

In this forest the Consul's brother lies hopelessly drunk; the Consul's wife is trampled to death by horses. Yet the Consul crawls through, he sees "The Lighthouse in the Storm;" but the lighthouse is *El Farolito*, the infamous *cantina* whose dark labyrinthine passages lead under the volcano.

In the last scene, as the Consul, drunk, dies from a gratuitous pistol bullet, he reflects on his native Kashmiri mountains, dreams of climbing the volcano Popocatepetl, but falls screaming into the *barranca*. And in one of the most telling last lines of any modern novel, "Someone threw a dead dog after him down the ravine."[67] Psychic journey's end.

Home

> Blue mountains with snow and blue cold rough water,
> A wild sky full of stars at rising
> And Venus and the gibbous moon at sunrise,
> Gulls following a motorboat against the wind,
> Trees with their branches rooted in air —
> Sitting in the sun at noon with the furiously
> Smoking shadow of the shack chimney —
> Eagles drive downwind in one,
> Terns blow backward,
> A new kind of tobacco at eleven,
> And my love returning on the four o'clock bus
> — My God, why have you given this to us?
>
> M. Lowry, "Happiness"

"Life is a ceaseless journey home."[68] Lowry's life, as that of his protagonists, was the ceaseless quest of the homeless for a home-place. London, Paris, Spain, Mexico, New York, Los Angeles, and least of all Liverpool, none of these could be home for Lowry. For a while home was personified in the novelist Conrad Aiken, mentor and father-figure to Lowry. But it was only with the unexpected trip to Vancouver that Lowry found, in Dollarton, a place he could call home. Throughout his letters and his travelogue-novels after 1940 Dollarton ("Eridanus") is called "home." In Europe in the late 1940s Lowry is hospitalized, alcoholic, insane. Back in the rickety beach shack in the squatter settlement of Eridanus he recovers remarkably.

The simple life of virtue envisioned by the Consul in *Volcano* was realised by Lowry in the 1940s. Across the water lay the ever-present evils of Enochvilleport (Vancouver), but in Eridanus the enclosing forest kept out civilization, one's neighbours were rustic fishermen, one took "The Forest Path to the Spring" for water, one met cougars in the forest and spoke to them:

> Far aloft gently swayed the mastheads of the trees: pines, maples, cedars, hemlocks, alders ... Beyond, going toward the spring, through the trees, range beyond celestial range, crowded the mountains, snow peaked for most of the year. At dusk they were violet, and frequently they looked on fire, the white fire of the mist.[69]

Lowry was acutely sensible of the landscape of Eridanus, of its diurnal and seasonal rhythms. Summer was, of course, idyllic, but even:

> The wintry landscape could be beautiful on these rare short days of sunlight and frost flowers, with crystal casing on the slender branches of birches and vine-leaved maples, diamond drops on the tassels of spruces, and the bright frosted foliage of the evergreens.[70]

And Eridanus was a settlement of the forest, and of the water; it was a natural, organic growth;

> ... everything in Eridanus ... seemed made out of everything else, without the necessity of making anyone else suffer for its possession: the roofs were of hand-split cedar shakes, the piles of pine, the boats of cedar and vine-leaved maple. Cedar and fir went up in chimneys and the smoke went back to heaven.[71]

There was no hatred, no one locked his door, no one spoke meanly. The neighbours' "little cabins were shrines of their own integrity and independence."[72] Here above all, one could forget civilization and get back to the existential reality of the things themselves: rocks, water, trees, wind, stars.

Yet even here Lowry felt sometimes a vague unease, an urge to become a travelling man, even a sailor, once again. And this despite his experiential knowledge of the falsity of the sailor's life, where the engine's rhythm sings, "You'll soon be home ..." and yet, stepping on shore, he finds himself homeless. For the sailor, indeed, home life was "reduced to a hip-bath with your wife on the kitchen mat every eighteen months, that was the sea."[73]

Three novel-travelogues depict the destructive point:counterpoint of home:journey, eutopia:dystopia, Paradiso:Purgatorio, eden:civilization, Eridanus:not-Eridanus that became the dialectic of Lowry's life during the period 1940-1954. *Dark As The Grave*[74] recounts a return journey from Canada to Mexico, from eutopia to dystopia, with recurrent reflections on the purgatory that is travel (p. 94) and constant wishes to return to the paradise that is Eridanus. Sigbjorn regularly reflects on home, "the little new still-unfinished house below, the cedar tree, the foreshortened pier, their boat hoisted up and overturned on the platform for safety during their absence ..." (p. 81), a home that the Wildernesses themselves had built (p. 55).

Hear Us O Lord[75] is an elegantly structured epitome of the chief Lowryan themes. In the first stories the protagonist sadly leaves Dollarton and passes through the Panama. The three central tales, set in Italy, depict the reflections of the thinking tourist. In Pompeii the character Fairhaven (which is what he is seeking) looks without seeing, reflects upon "the malice of travellers, even the sense of tragedy that must come over them sometimes at their lack of relation to their environment."(p. 177)

The tourist is always inauthentic, the traveller often so:

> The traveller has worked long hours and exchanged good money for this. And what is this? This, pre-eminently, is where you don't belong ... and behind you, thousands of miles away, it is as if you could hear your own real life plunging to its doom (p. 177).

Real life, of course, is lived on Dollarton beach. The last two stories in the novel, and in particular the novella "The Forest Path To The Spring" are a paean of praise for home, for Eridanus. The complete structure, then, is the familiar one of withdrawal and return, and the withdrawal, as in *Lunar Caustic*, is usually a journey into purgatory.

October Ferry,[76] a rather poor novel in a literary sense, is nonetheless one of the most valuable for its clear exposition of the recurrent Lowryan themes. The notions of home, journey, exile, and the quest for a new home are present throughout. This is a novel of Paradise Lost. The Llewelyns "like love and wisdom, had no home." (p. 5) Homes burn down around them. They are evicted, "evicted out of exile" even (p. 4). They consider the homes of others, for many of whom "home from home" is merely the men's side of that truly awful institution, the Canadian beer parlour (p. 43). Postwar newspapers ironically speak of "Work to the Workers, Homes to the Homeless." (p. 46) Llewelyn is looking for a place that says "not 'I am yours', but 'You are mine.'" (p. 51)

The Llewelyns are about to be evicted from their squatter shack at Eridanus. They celebrate the winds and tides of their home. Although the shack seemed impermanent, "antiquity of mountains, forest, and sea, conspired on every hand to reassure and protect them, as with the qualities of their own seeming permanence." (p. 79) Between the cabin and the Llewelyns was "a complete symbiosis. They didn't live in it ... they wore it like a shell." (p. 80)

But it was "Back to nature, yet not all the way. Rousseau with a battery radio." (p. 154) Suburbia creeps towards Eridanus. The Rosslyn Park Real Estate and Development Company is erecting the suburb of Dark Rosslyn with "Scenic view lots. Approved for National Housing Loans. Cash or Terms." Dollarton has become, for Lowry, *Dolorton*. His safe haven is threatened. The Dollarton squatter shacks, embodiment of "an indefinable goodness, even greatness," are to be swept away by a tide of progress. "A suburban dementia launched itself at them;" it was as if "they want to turn this whole place into a vast bloody great Black Country, a Lancashire ... of the Pacific Northwest."[77] It was time to go. And in 1954, their Eridanus reduced to a tiny oasis of forest surrounded by suburbs and oil refineries, the Lowrys decided to leave.

The Lowrys (the Wildernesses, the Llewelyns) were evicted not by legal order but by civilization and progress. At Vancouver Airport Lowry cried "I'm afraid to leave. I'm afraid we'll never come back."[78] Expelled from his eden, the Wandering Jew nursed memories of Eridanus and especially of the rickety pier that he himself had built. "To me ... childish though it may seem there is the pier, which we built, which I cannot imagine myself living without", he wrote from Sussex.[79] That the pier had been swept away by storms was kept from Lowry for some time. On hearing of its fate he was "broken-hearted".[80] He died, by misadventure/suicide, the following year.

CONCLUSION

The Lowryan opus, confirms the value of the antinomies set out in general terms by humanist geographers. Home and journey are indeed fundamental geographic poles, insideness and outsideness are existential conditions of the greatest import. As with Williams and the Whites, we confirm that for sensitive beings the city is likely to be dystopia and unmanageable, whereas a life closer to nature may be eutopia.[81] This was to be expected.

But Lowry's contribution, in life as in work, lies in taking these conditions to their extremes. Lowry and his protagonists are rarely at rest; home is ideal, but journey is real. His work is that of the existential outsider who, briefly admitted to a form of insideness, is nevertheless eventually cast out. Sea:land is indeed a primary dialectic, but in Lowry the balance is lost. Land becomes dystopic, especially in cities, and can be endured only on the sea edge. By the sea one is liberated; at sea one is free. Yet Lowry himself realized that this again was an ideal, only a temporary condition; land, with all its ambivalence, must be reached eventually.

Home is often equated with self; Kahlil Gibran tells us; "Your home is your larger body."[82] Yet with Lowry the identification with home becomes so extreme that survival without land, shack, and pier is fundamentally impaired. This, again, was clear to Lowry himself. In *October Ferry* Llewelyn's wife and father-in-law refer to his attachment to Eridanus as "insane" and "pathological". To which Llewelyn can only agree, yet stating: "not to be attached ... would be more pathological still."[83] Less intense, but equally real attachments to home are felt by many individuals and groups. Yet politicians, planners, and developers, unheeding of the humanist planning research which began in Boston's West End, neither know or care that grieving for a lost home may be fatal.[84]

Lowry's work also brings into focus the two types of journeying, of body and of mind, the former often symbolic of the latter. Despite the recent trend towards a consideration of process, geographers still tend to take a static view of the world. Space and place are emphasized; when considered, journeys are always considered as place-related, rarely as events in themselves. We are slow to consider life-journeys, or the long-term journeying of drop-outs, exiles, and expeditionaries, preferring to concentrate on commuters and tourists (who, in the geography of tourism, seem to exist only to be managed).

Journeys of life, inner psychic journeys, allegorical journeys, these are not explored. And in our emphasis on at-homeness, rootedness, and place we are missing a great deal of the point of even nineteenth century regional novels, as Middleton has shown.[85] As for the modern mid-twentieth century novel, it reflects the growing placelessness of our civilization. As yet, however, geographers' studying of imaginative literature seems loathe to leave the known parameters of life before 1914.

More generally, Lowry's work has a number implications. It has become normal in humanist geography to look at landscapes in terms of antinomies, such as city:countryside in Britain and city:wilderness in North America. Tuan specifically provides us with a city:country:wilderness model for understanding landscape and life.[86] But such models are essentially static, whereas the relationship of city, countryside, and wilderness is always a dynamic one. In Lowry's case the linkage, for the individual, is by means of the journey.

More profoundly, the use of polarities such as city:wilderness is erroneous and reactionary in terms of power relationships. Wilden has demonstrated how the use in Canada of such dialectics as white:black and capital:labour obscures the reality of the relationship between the antagonists.[87] For "lock-out" is generally more powerful than "strike" and "nigger" a more powerful epithet than "honky". A more truthful diagrammatic rendering of these relationships would place the dominant group clearly over the subordinate, as:

$$\frac{\text{white}}{\text{black}} \quad \text{and} \quad \frac{\text{capital}}{\text{labour}}$$

and, in the case of the urban Leviathan:

$$\frac{\text{city}}{\frac{\text{countryside}}{\text{wilderness}}}$$

Finally, Lowry's work is valuable in that it brings to our attention the ethical dimension of landscape character. As Relph has shown, sensitive persons may rail at the aesthetic nightmare that characterizes the "urban strip" on the fringes of North American cities, but utterly fail to appreciate the barbaric treatment of living creatures which supports the garish fast-food outlets so prevalent in this environment.[88] Lowry asks who is responsible for the tasteless urban scene and for the wreckage of the natural environment around it.[89] The Pogo comic strip supplied the existentialist answer decades ago: "We have seen the enemy, and he is us." In the long run, we get the environments we deserve; a rather chilling thought, perhaps, for contemporary Canadians.

ENVOI

Which brings me to the impersonal nature of humanist geography. It has become a cliche for humanist geographers to end their essays with statements to the effect that such studies will sensitize the reader to the nature of the world, help the student be a better citizen of both world and place, and generally increase the burden of awareness. No evidence is ever provided to support these pious platitudes.

Moreover, although humanistic geography purports to investigate the existential situation of individuals as well as groups and humanity at large, humanist geographers never seem to have come to grips with *themselves* as exemplifiers of phenomenological reality. One looks in vain in the writings of Lowenthal, Tuan and the like for some evidence of the writer's lived experience, but we find only the experiences of *others*. For me, an existential, phenomenological approach must deal with *felt* experience, and my felt experience as much as that of others. So I must now report that I feel close to Malcolm Lowry, not in terms of his psychic problems or his literary genius, but in the sense of: (1) his homelessness, and search for a home-place; (2) his insatiable urge to travel; (3) his experience as a temporary resident in Cuernavaca, Mexico, and as a permanent resident of southwestern British Columbia; and (4) his happiness, despite exile from Britain, in a seashore cabin off the British Columbia mainland.

And with the life = work example of Lowry before me, I also ask: Why do geographers study the things they study? Not surely because of overbearing graduate supervisors who neatly invested one with a rut, but surely, for some, because of deeper needs? I can only answer for myself. Papers on company towns in remote places, on literature, on home; books on urban history, canal ports, environment and behaviour, Easter Island;

current work on autogeobiography; all these have something to say about my childhood, expectations, need to travel, love of the exotic, need for home in exile, need to understand my life experiences.

My work, like that of Walt Whitman or the eastern churchman Gregory Nazianzen, is a "Song of Myself". It may be of interest to others, it may advance scholarship, but it is most valuable to me. Brought up in a tiny village, my early interest in urban geography was a defensive act, a need to understand the city, which from childhood has been an alien place. Understanding, of course, may bring toleration, but in this case has simply confirmed that my childhood self was right. Can one, like Lowry, be allergic to civilization? This paper, like most of Lowry's books, was written on the semi-wilderness coast of British Columbia, among the eagles and seals.

REFERENCES

1. GILBERT, E.W., "The Idea of the Region", *Geography*, 45, 1960, pp. 157-175.

2. SALTER, C.L. and LLOYD, W.J., *Landscape in Literature*. Washington, D.C.: Association of American Geographers Resource Paper 76-3.

3. TUAN, Y-F., *Space and Place*. Minneapolis: University of Minnesota, 1977.

4. PORTEOUS, J.D., "Literature and Humanistic Geography", *Area*, 17, 1985, pp. 117-122.

5. EYLES, J., *Senses of Place*. Warrington: Silverbrook Press, 1985.

6. GOLD, J. and BURGESS, J., *Valued Environments*. London: Methuen, 1982.

7. PORTEOUS, J.D., "Surname Geography: A Study of the Mell Family Name c. 1538-1980", *Transactions, Institute of British Geographers* NS7, 1982, pp. 395-418.

8. JOHNSTON, N., *You Can Go Home Again*. New York: Doubleday, 1982.

9. These notions are explored, *inter alia*, in POCOCK, D.C.D. (ed.), *Humanistic Geography and Literature*. London: Croom Helm, 1981; RELPH, E., *Place and Placelessness*. London: Pion, 1976; TUAN, 1977, *op. cit.*

10. PORTEOUS, J.D., "Home: The Territorial Core", *Geographical Review*, 66, 1976, pp. 383-390.

11. PORTEOUS, *op. cit.* (reference 4)

12. PORTEOUS, J.D., "Inscape: Landscapes of the Mind in the Mexican and British Columbian Novels of Malcolm Lowry", *Canadian Geographer*, 30, 1986, pp. 123-131.

13. For a fuller discussion, see DAY, D., *Malcolm Lowry*, New York: Dell, 1973.

14. BREIT, H. and LOWRY, M.B., (eds.), *Selected Letters of Malcolm Lowry*. Philadelphia: Lippincott, 1965.

15. WOODCOCK, G., "Under Seymour Mountain", *Canadian Literature*, 8, 1961, p. 4.

16. NEW, W.H., "Lowry's Reading", pp. 125-132, in WOODCOCK, R. (ed.), *Malcolm Lowry: The Man and His Work*. Vancouver: U.B.C. Press, 1971, p. 131.

17. LOWRY, M., *Ultramarine*. London: Penguin, 1974.

18. LOWRY, M., *Under The Volcano*. London: Penguin, 1963.

19. LOWRY, M., *Lunar Caustic*. London: Cape, 1968.

20. LOWRY, M., *October Ferry To Gabriola*. New York: Plume, 1970.

21. LOWRY, M., *Hear Us O Lord From Heaven Thy Dwelling Place*. London: Penguin, 1969.

22. DAY, *op. cit.*, p. 342.

23. LOWRY, M., *Dark As The Grave Wherein My Friend Is Laid*. London: Penguin, 1972, p. 209.

24. LOWRY, *Dark As The Grave. op. cit.*, p. 84.

25. PORTEOUS, J.D., *Environment and Behaviour: Planning and Everyday Urban Life*. Reading, Mass.: Addison-Wesley, 1977.

26. LOWRY, *Hear Us O Lord. op. cit.*, p. 129.

27. LOWRY, *Dark As The Grave. op. cit.*, p. 215.

28. LOWRY, *Volcano. op. cit.*, p. 41.

29. TUAN, Y-F, "The City: Its Distance from Nature", *Geographical Review*, 68, 1978, pp.1-12.

30. LOWRY, *Dark As The Grave. op. cit.*, p. 115.

31. LOWRY, *Hear Us O Lord. op. cit.*, p. 215.

32. LOWRY, *October Ferry*. *op. cit.*, p. 181.

33. LOWRY, *Hear Us O Lord*. *op. cit.*, p. 262.

34. *Ibid.*, p. 98.

35. *Ibid.*, p. 284.

36. LOWRY, *October Ferry*. *op. cit.*, p. 159.

37. LOWRY, *Dark As The Grave*. *op. cit.*, p. 78.

38. LOWRY, *Hear Us O Lord*. *op. cit.*, p. 14.

39. *Ibid.*, p. 15.

40. *Ibid.*, p. 14.

41. *Ibid.*, p. 15.

42. *Ibid.*, p. 261.

43. TUAN, Y-F., *Landscapes of Fear*. New York: Pantheon, 1979.

44. LOWRY, *Dark As The Grave*. *op. cit.*, p. 51.

45. LOWRY, *Hear Us O Lord*. *op. cit.*, p. 179.

46. LOWRY, *October Ferry*. *op. cit.*, p. 229.

47. A collage of quotations from LOWRY, *October Ferry*. *op. cit.*, p. 64 and *Hear Us O Lord*. *op. cit.*, pp. 227, 238, 276.

48. LOWRY, *Hear Us O Lord*. *op. cit.*, p. 207.

49. *Ibid.*, pp. 202-207.

50. *Ibid.*, p. 205.

51. LOWRY, *October Ferry*. *op. cit.*, p. 202.

52. LOWRY, *Hear Us O Lord*. *op. cit.*, p. 279.

53. LOWRY, *October Ferry*. *op. cit.*, p. 43.

54. LOWRY, *Hear Us O Lord. op. cit.*, p. 125.

55. *Ibid.*, p. 188.

56. LOWRY, *Dark As The Grave. op. cit.*, pp. 171.

57. *Ibid.*, p. 188.

58. LOWRY, *Dark As The Grave. op. cit.*, pp. 36, 42.

59. LOWRY, *October Ferry. op. cit.*, p. 192.

60. LOWRY, *Hear Us O Lord. op. cit.*, p. 198.

61. DAY, *op. cit.*, pp. 124, 143.

62. FUSSELL, P., *Abroad.* New York: Oxford University Press, 1980.

63. DAY, *op. cit.*, p. 344.

64. LOWRY, *Lunar Caustic. op. cit.*, p. 17.

65. DAY, *op. cit.*, p. 344.

66. *Ibid.*, p. 306.

67. LOWRY, *Volcano. op. cit.*, p. 375.

68. PORTEOUS, 1976, *op. cit.*

69. LOWRY, *Hear Us O Lord. op. cit.*, p. 216.

70. *Ibid.*, p. 253.

71. *Ibid.*, p. 248.

72. *Ibid.*, p. 247.

73. LOWRY, *Volcano. op. cit.*, p. 197.

74. LOWRY, *Dark As The Grave. op. cit.*

75. LOWRY, *Hear Us O Lord. op. cit.*

76. LOWRY, *October Ferry. op. cit.*

77. *Ibid.*, p. 201.

78. DAY, *op. cit.*, p. 423.

79. *Ibid.*, p. 42.

80. *Ibid.*, p. 43.

81. WILLIAMS, R., *The Country and The City*. London: Paladin, 1975; WHITE, M. and L., *The Intellectual Versus the City*. Cambridge, Mass.: Harvard University Press, 1962.

82. GIBRAN, K., *The Prophet*. New York: Knopf, 1976.

83. LOWRY, *October Ferry. op. cit.*, p. 199

84. PORTEOUS, 1977, *op. cit.*

85. MIDDLETON, C.A., "Roots and Rootlessness: An Exploration of the Concept of the Life and Novels of George Eliot", in POCOCK (ed.), *op. cit.*, pp. 101-120.

86. TUAN, Y-F., *Man and Nature*. Washington, D.C.: Association of American Geographers Resource Paper 10, 1971.

87. WILDEN, A., *The Imaginary Canadian*. Vancouver: Pulp Press, 1980.

88. RELPH, E., *Rational Landscapes and Humanistic Geography*. London: Croom Helm, 1981.

89. LOWRY, *October Ferry. op. cit.*, p. 217.

Hagan Creek Valley (**PLATE 11**) and Mt. Work (**PLATE 12**) are only a couple of kilometres apart on the Saanich Peninsula, British Columbia, but have very different colour characteristics. ►

5 COLOUR AND LANDSCAPE

Colin Wood

INTRODUCTION

The arrangement of physical and cultural features on the surface of the earth which we know as 'landscape' is a complex idea which embraces both measurable phenomena and subjective tastes and values. Colour, or the variation in light energy wavelength, is an important element of the visual perception and appreciation of value and beauty in landscape. Together with landscape, which is itself comprised of many things, the attribute of colouration can be measured objectively and appreciated in aesthetic terms.[1]

This chapter examines the characteristics of colour as a facet of landscape with a general purpose of elucidating its characteristics, meaning and significance. First, the hypothesized relations between cause, sensation and appreciation are drawn together as a simple schema. Attention then turns to the problem of description and measurement. One aspect of the link between people and landscape colour is explored through a review of situations where landscape colours have been significantly changed. The chapter concludes with some commentary on the meaning of colour and landscape. It is not the purpose of this discussion to examine the appreciation of colour in respect to aggregation of individual perceptions and tastes, a topic beyond the scope of this essay, but nevertheless one of major importance. In itself however, the chapter is an individual appreciation of landscape colour and has an objective of enhancing awareness of its geographical attributes.

PHYSICAL CAUSE: LIGHT SOURCE, OBJECT AND RECEPTORS

Colour

Colours are the result of light energy exciting the pigmentation of the atomic structure of objects which act as "a minute machine, capable of absorbing and thus being excited or energized by electromagnetic radiation in the form of light which later loses in one or a variety of ways, fluorescence, phosphorescence, heat or transfer to some other molecule."[2] The interaction between light energy and a surface responding in the above manner results in a particular part of the light spectrum being excited whether a human sees it or not. The colour which may occur is also a function of the object's shape, whether organic or incorgnaic and its surface, whether smooth or uneven. The colour which may be observed will be a function of the condition of the light energy and the relative location of object and observer. Further variation can occur according to time of day, season and geographic location. The viewing situation can be a permutation of altitude, relief, meteorological conditions and human activities such as air pollution.

Receptors

The ability to see, perceive, identify and differentiate the colour of objects is dependent initially on the receptors that is, the whole visual system including the eye. Isaac Newton discovered over 300 years ago that white light is composed of the seven basic primary colours and a mixture of these gives the great range around us. The human focal system is sufficiently well developed for normal healthy individuals to distingish at least 350,000 "just noticeable differences" (JND) in colour.[3] A specific colour is a combination of hue, brightness and saturation. How these colours are perceived is conceptualized as either:

1. the reception of three primary hues by cones in the eye (Young, Helmholtz theory)
2. the reception of four lines by pairs of receptors (the opponent colours or Hurrich, Jameson theory).[4]

Not everyone has complete colour vision due to the occurrence of disease or genetic abnormalities. Statistics compiled by Kalmus (1965)[5] suggest that colour blindness rates are higher among European and North Americans (7 to 10 percent) and low among primitive peoples (2 to 3 percent). Post (1962)[6] hypothesizes that variations of this type are due to genetic selection and the length of time a society has evolved from a hunting economy which would, it is argued, relax the bar on genes with defective colour vision. Presumably, after many generations of a civilization which no longer needs to detect prey and now subsists on refined foods, colour blindness could become the norm for urban society.

Consequently, as light of a particular condition strikes an object, which because of its colour property will have a specific permutation of hue, brightness and saturation, it will cause a stimulus in observers. The perception, cognition and meaning of an object's colouration can in turn be related to the observer's physiological condition, intellect, learning experience and general cultural background. Thus, the general relationships between light source, colour property of the object, transmission effect due to intervening atmospheric conditions and condition of the observer are set out in Figure 1,5. The diagram is a simple description of the components of the landscape colour system according to the sense of vision, based on the simple triad of light, object and observer. Research in several disciplines has expanded knowledge of the detailed, disaggregated components of this system. However, there are few systematic generalizations of landscape colour in a geographic context although there are innumerable individual descriptions by authors, poets and painters.[7] The following generalizations are based on the author's observations and are an attempt at a simple initial identification of landscape colour conditions from a geographic perspective.

Landscape Colour

Landscapes are generally composed of several dominant features, topographic relief, vegetation, settlements and so on. Their colour can vary significantly so that the landscape is in effect a combination of colours, the colours being, of course, indicators of the component features. Other indicators are size, shape, proximity, all significant according to recognition by the observer. Yet, unlike shape and size of objects there are perhaps few other phenomena that humans observe in their awareness of the environment that are as ephemeral as colour. The particular permutation of hue, brightness and saturation are unique to hour, day, season and atmospheric conditions with the likelihood that for certain geographic locations no one permutation is ever exactly repeated. Nevertheless, some generalizations are possible and necessary otherwise there would be no recognition in the colour content of objects.

LANDSCAPE COLOUR ASSOCIATIONS

Some Generalities

There is agreement that certain landscapes have a particular colour association, for example, near and middle ground hills in temperate latitudes tend to be dark green due to their vegetation cover while the sea tends to be a blue-green grey. However, while a spectacular sunset may make the same hills appear fiery red due to the observation and peculiar light conditions of that moment, the observer knows from experience that such colours are ephemeral. Similarly, a rare thunderstorm may turn the

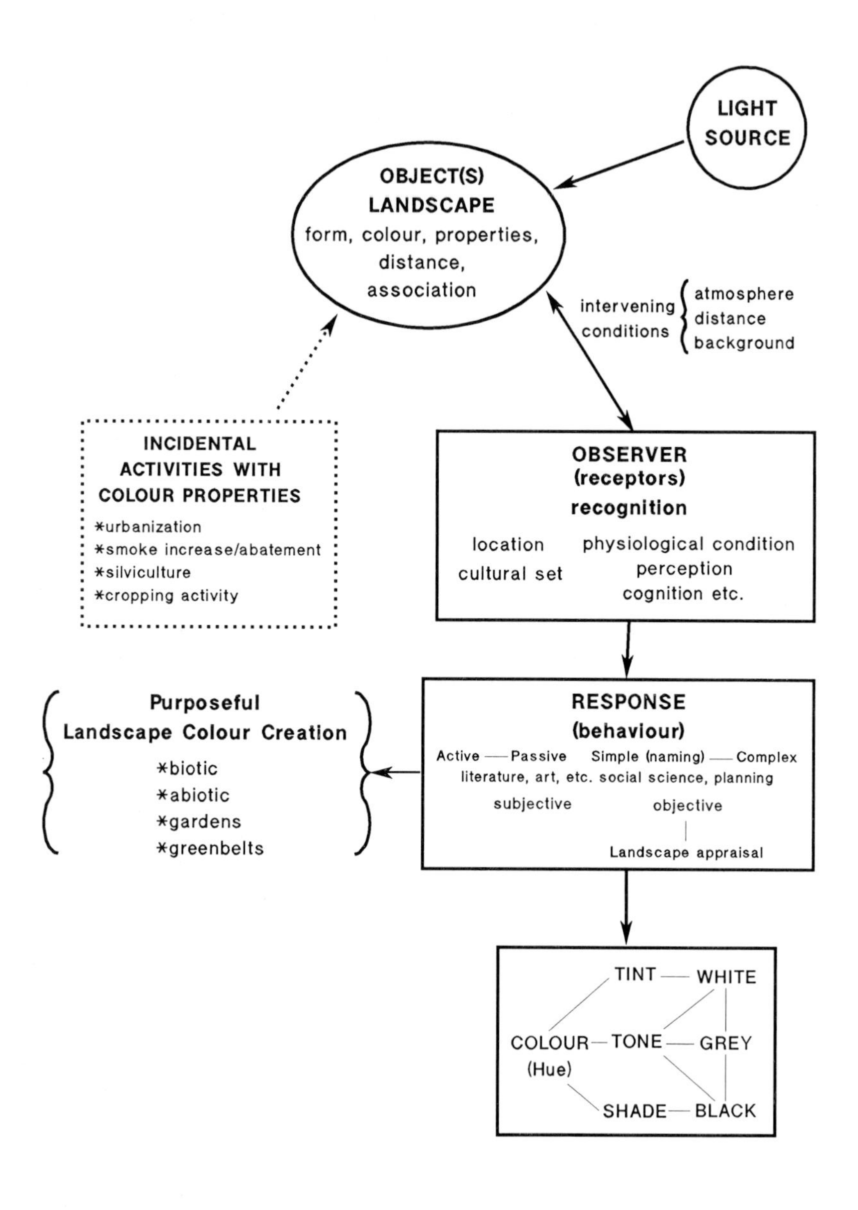

FIGURE 1,5 The Interaction System Between People and Landscape Colour Conditions

grey or yellow desert or semi-arid landscape into a profusion of colours which lasts a few days. The 'permanent or expected' colour association is generally prevalent at a regional scale reflecting the particular combination of physical elements and the observation conditions, thus suggesting a broad regional condition which can be designated *colour dominants.* Furthermore, it is probable that the inhabitants (observers) of that region are, through habituation, visually attuned to the usual colour palette of that region.

An interesting aspect of the colour dominants of regions is that artists of all kinds are as much motivated to capture the unusual optical conditions of short-lived momentary or seasonal bright colours as they are the general regional colour dominants. Perhaps because during unusual observation conditions people pay more attention to the landscape and hence the drama of the creations captured by the artist. Consequently, a geographic variation in colour dominants can be differentiated on a regional and perhaps even a local basis, constituting a basic condition of landscape colour.

An important geographic regularity is the effect of distance between the object and the observer together with the influence of intervening moisture or dust on viewing conditions. Middleton calculates that according to the optical properties of air and light, dark objects look blue in the distance the more saturated the air, while in clearer or drier air, objects of bright positive contrast look pale orange (snow-capped peaks).

> ... it is really the blue of the sky ... which lies in front of the distant hill, though the difference in luminance may hinder the recognition of its identity. At 30 kilometres the hill is less than 30 percent as bright as the horizon.[8]

This regularity has importance in distance perception and for portraying landscape depth in paintings while 'blue hills', 'blue ridges', 'cairn gorm' all demonstrate this regularity in perception in placenames. The significance of colouring objects brightly in foregrounds and colouring them in muted blues and greys in backgrounds were important steps in the evolution of landscape art and the artist's ability to convey the impression of three-dimensional landscapes in a two-dimensional medium.

Many urban residents are aware of the effect of smoke and emissions on colour visiblity. When travelling to high latitudes observers are frequently startled by the clarity and colours of the landscape, especially when the moisture and dust content of the atmosphere happens to be low. In regions where moisture levels are high and distant objects fade into a damp grey/blue haze the monochrome background effect may be accentuated by bright green vegetation in the foreground.

An intriguing characteristic of landscape appreciation is the special awareness and attention given by artists to short-lived or unusual atmospheric conditions, for example, the dramatic colours which can occur at sunset or during and after storms, rainbows and rain showers, these are noticed as unusual and are frequently the subject of interpretation. Indeed, one *tromp l'oeil* is to portray monochrome temperate landscapes in variegated sunlight — conditions with shafts of sunlight creating dramatic light and colour contrasts against subdued backgrounds. Thus, even rather ordinary landscapes may become extraordinary at sunset or during a summer thunderstorm.

Artists are particularly sensitive to light and atmospheric conditions, striving to capture the unique effect of certain observation conditions of geographic regions (the Canadian lakes, the Mediterranean) through their own creative endeavours. The Romantic period saw the use of colour to appeal to the emotions. Other schools of style have concentrated on certain colours or optical effects. Some artists painted with their back to the light (Mauve) while others (Maris) painted towards the source of light. Many artists have sought to portray the particular light conditions of certain regions particularly where colours are bright and dramatic contrasts are present.[9]

As relief becomes more rugged and increases in altitude, the impact of light and shadow effects will increase the variety of colouration compared with plane surfaces. Such conditions will be further enhanced with the presence of water bodies which can reflect light and modify colours of other objects. The enriched light and colour conditions by rivers, lakes and the seashore are consequently attractive both to artists and as areas for residence. Other elements of the objects of observation also contribute to the colour of the landscape, particularly vegetation and exposed rock or soils. Vegetation, by nature of its biological structure generally tends to be green transforming to blue/greys with greater intervening observation distance. Evergreen vegetation tends to have greater uniformity in colour while deciduous vegetation is often variegated with pronounced seasonal alteration in colours. Desert landscapes, by contrast, have warm colours, yellows and browns, and occasionally bright colouration due to exposed rock surfaces, minerals such as iron oxides and red sandstones. Areas of water evaporation and salt precipitation give conditions of dazzling whites. Generally vegetation and rock or soils play an important role in the fore and middle ground perspective of landscape. Regions with greater variety of plant types and seasonal changes in vegetation will tend to show greater variety in colours, whereas coniferous, prairie and desert landscapes tend to be relatively stable in colouration — with the connotation of monotony.

Observers

Without extensive literature available on aggregated observer perceptions' of landscape colour the following statements should be treated with considerable caution, nevertheless the author feels that they have some generalized validity. Most observers see landscape colours in the horizontal plane which can be roughly categorized as major bands of fore (near), middle and background. With greater use of air transport, however, more people are now seeing landscapes in the vertical plane. Observers are naturally most familiar with the horizontal plane and the dominant colouration of their landscape of usual habitation.

The preference for certain regional landscapes is discussed in the literature on landscape taste but there are few references to the significance of colour. It is assumed that generally observers prefer vegetated landscapes, to desert landscapes and variety to monochromes. Observers in cool monochrome regions are stimulated by bright colours and endeavour to seek them out as evidenced by trips to — bulb growing and fruit regions in spring, deciduous regions in the fall — and, of course, through the propagation of bright flowering plants. Generally bright colours are only observable in the foreground since their effect diminishes appreciably with distance. The exception are large, white objects (snow-capped mountains and limestone cliffs) which may be visible for considerable distances. The appreciation of colour in landscape is particularly heightened in artists, poets and authors. However, this is not to say the person in the street is not aware of landscape colours, rather, he or she is less sensitive for example in the sense of having knowledge of colour palettes and less trained in communicating an interpretation and understanding of them.

RECOGNITION, DESCRIPTION AND MEASUREMENT OF COLOUR AND LANDSCAPE

Until the advent of modern methods of colour photography, the description and measurement of landscape colour has been both rudimentary and technically difficult. To start with, in spoken and written expression there is a paucity of colour terms despite the very large number of colours that can be detected by the eye.[10] The two main technical problems of recording colours have consisted of developing pigment production to duplicate observations and recording colours in a fast enough time frame before they change. Constrained in this manner artists had therefore to strive for an expression that was in essence an approximation of an ephemeral condition. The development of accurate colour printing, colour classification systems such as Munsel, colour photography, film and television has made a consistent form of recording, measurement and interpretation of landscape colour possible.

The earliest recognition of colour in a geographic context predates modern technology by centuries and can be found in place names and place/colour associations. Ancient upper Egypt was represented by a white crown and Lower Egypt a red one — when the two were united, the two were combined but it was then found to be cumbersome so a blue crown was then used.[11] In local and regional place and topographic feature names colour is used both as a pre and suffix. In Bartholomew's *Gazetteer of the British Isles*, with 90,000 names, one percent have a colour attachment.[12] In Canada, where settlements took place mainly after the standardization of spelling, approximately two percent have a colour attachment (of a total of 250,000 place names).[13] Despite the difference in place and time of settlement there is a marked similarity in the frequency of colour use (Table 1,5). Black and white although not colours *per se* are used perhaps because places or features of very dark or bright contrast do not occur frequently and can therefore be given the colour/place association. From recognition we can turn to description and measurement.

Representations and descriptions of landscape colour are found in landscape painting and literature. They all demonstrate the perception by elite observers, the artist/author, and their technical skill at conveying the image to the observer or reader. Each impression is a combination of the individual author or artist's perception of reality and its rendering according to conventions of taste and style. Content analyses of literature show that black and white are the most frequently used 'colour' terms, followed by red, green and blue. The categorization of colour use in landscape art is more difficult. Certain groups of artists or schools tended to use certain palettes of colours in their compositions. However, until the modernists, the intention was a certain degree of realism between an objects' colour and its representation.

Before the development of modern colour photography, painting and literature were the main focus of recording, describing and interpreting landscape colours. Perhaps because of the individuality and creativity involved (poetic license) in the recording process, generalizations could not be possible other than 'the sea is blue', 'the grass is green' and similar statements.

MEASUREMENT: AN EXAMPLE

The development of colour classifications such as the Munsel system and colour photography have now made possible the consistent recording and measurement of landscape colours. Consequently, the recording and interpretation of regional colour dominants becomes feasible. There are

Table 1,5 PLACE NAMES USING COLOUR THEMES

U.K.	Black	220	Green	210	White	190	Red	78	Brown	34	Gold	33
Canada	Black	909	White	593	Green	566	Red	459	Brown	163	Blue	163

three main factors which need to be considered when recording and measuring landscape colour: first, the scale of detail, second, the perspective, whether horizontal, oblique or vertical and third, the time sample, that is, whether a photograph taken at 1/125 second or a year's recording of colour dominants is the objective.

1. *perspective*: the traditional perspective of landscape art and literature was mainly the single direction horizontal plane although cartography had vertical perspectives, with hand colouring, and occasionally 360° panoramas were executed. The development of aerial photography added a completely new dimension to interpretation. The vertical plane can provide more information per pixel than the horizontal, the latter being constrained by limitations of earth curvature, atmospheric moisture and relative obscuring of objects. Whereas the vertical may provide more information, the horizontal plane is the perspective most often seen by observers.
2. *scale*: clearly, the distance between object and observer, or the geographic scale of viewing, will dictate the degree of generalization involved. A larger scale will permit detailed colour recognition whereas a small scale will be dominated by monochromes. Generally bright colours are only observable at a large scale.
3. *time frame*: photographs record conditions at an instant whereas landscape colours are constantly changing. In developing a description of regional colour dominants, a compromise may be possible between the instantaneous photograph and a continuous record — giving generalization by season or month.

With these three conditions in mind, we can turn to a specific example namely the "normal" colour viewing conditions for the Saanich Peninsula at the southern tip of Vancouver Island, British Columbia. Figure 2,5 shows the "normal" or expected colour arrangements as seen in plan form (plane) — based on air photographs, taken on a clear sunny day during late spring early summer.

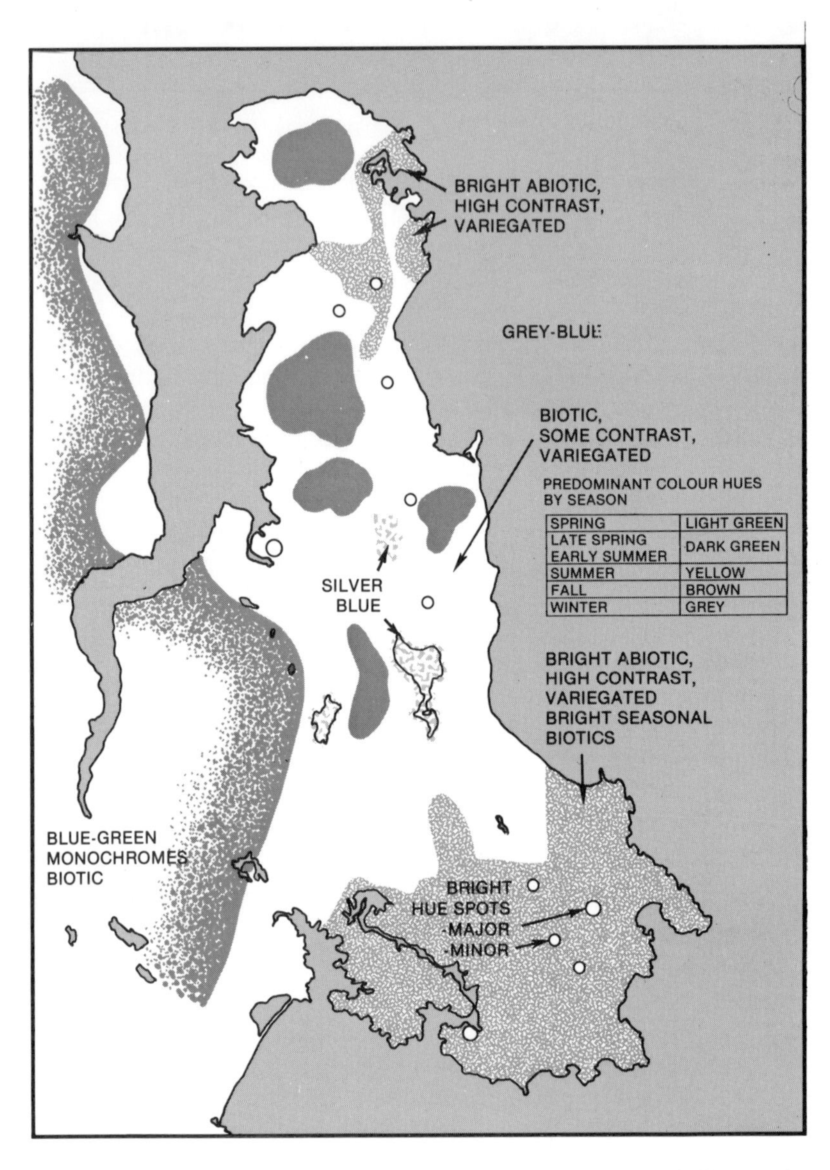

FIGURE 2,5 Normal Landscape Colours for the Saanich Peninsula

It can be seen that the region's land mass has a predominance of blue/green monochromes (evergreens), interspersed with seasonally variable deciduous and variegated cultivated areas. Bright hue colour spots exist where orchards bloom and in certain elevated locations, where broom (*Cytisus scoparius*) has bright yellow flowers in May and June. At this scale (1:125,000) individual bright colours are barely distinguishable. The temporal progression of colouration is noticeable on the cultivated land and the deciduous oak forest areas: elsewhere, the evergreen conifers remain blue/green monochromes throughout the year.

At ground level, from the horizontal perspective, the landscape is seen as bands of colour — roughly horizontal in arrangement (Figure 3,5). With increased distance the background colours diminish into blue-greens with the notable exception of background white, snow-capped peaks.

Middleground colours are seen as mixtures of greens, browns and sometimes yellows. Foreground colours depend on specific land use and the predominance and mix of particular species of vegetation.

For example, groves of arbutus (*Arbutus menziesii*) a broad-leaved evergreen tree are characterized by dark green leaves and bright amber or red limbs or light yellow florescence during the flowering period and red berries during late summer and fall. Individual foreground meadows may be a variety of hues depending on type of grass and wildflowers and drainage. The only bright coloured flowering native tree is the dogwood with white blossom visible up to about one kilometre. Native maples can turn bright yellows and orange yellow/pinks in the fall. The temporal progression of colour arrangement for specific locations is also shown.

Through recording landscape colour in this manner it is possible that a geographic information base can be developed which makes more meaningful its analysis, interpretation and appreciation. The information would be an important benchmark of landscape colour in the late 20th century. The significance of these two points are underlined in the final section which deals with the structural changes in landscape colour caused by human activity.

COLOURING THE LANDSCAPE

There are many links between people and the colour of the landscape. A major element is the capacity humans have for changing landscape colour. Significant actions which have modified the natural colours of the landscape can be related to the evolution and geographic migration of humans across the face of the earth. In the first instance, changes in colour of objects were incidental to other purposeful activity such as land clearing

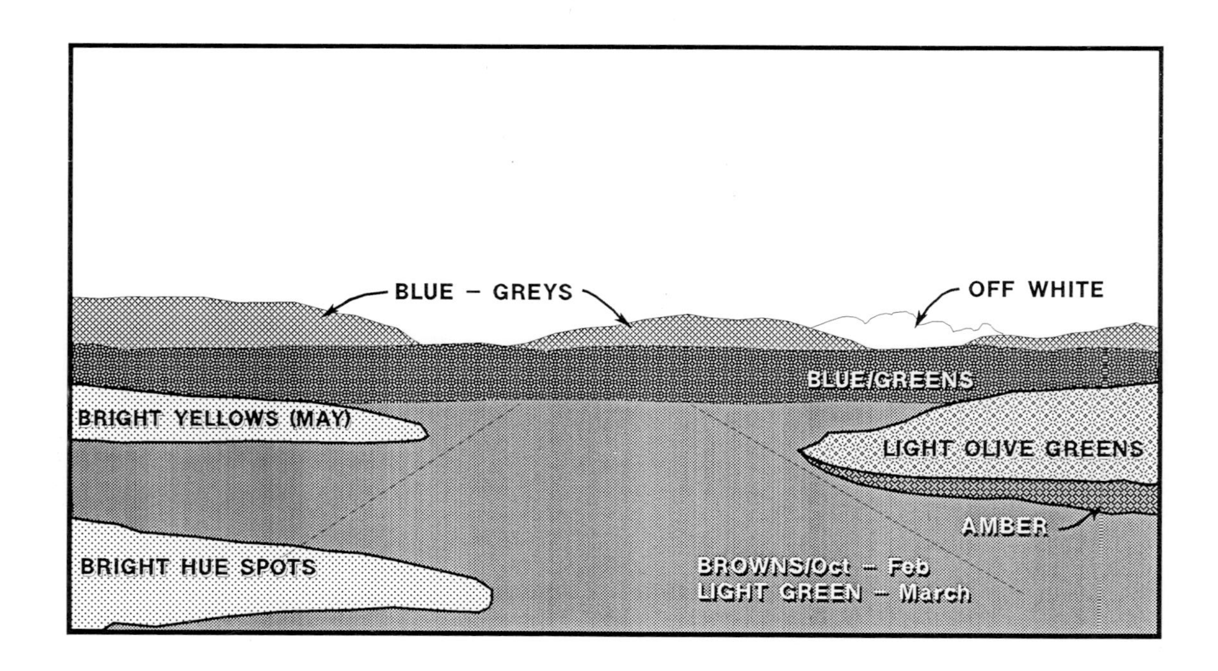

FIGURE 3,5 Colour Dominants Under Normal Viewing Conditions, Southern Vancouver Island, Horizontal Perspective

or irrigation. Recently there have been those practices which have a specific purpose in altering landscape colour. In addition, optical or viewing conditions have been altered through the presence of greater dust and smoke particles in the atmosphere particularly in the northern hemisphere. Occasionally alterations in atmospheric moisture content have also affected viewing conditions, for example through irrigation in semi-desert and desert regions.

Landscape colours have been changed by different degrees in many regions of the planet by human activities:

Activity	Colour Change
Land clearing: deforestation	Dark greens — light and variegated
Selection and cultivation of certain plants	Varied (plant specific)
Monoculture farming	Extensive monochromes
Irrigation and dry lands	Brown/yellows — greens
Reforestation with exotic species	Monochromes
Settlement and building construction	Colours to tones high contrast abiotics

Changes such as these have occurred since antiquity and are still happening today with forest clearing in the tropics (green to brown/variegated) more acreage given over to rapeseed (canola) production, (green to bright yellow) and the expansion of urban areas (the proverbial black top) to name a few. While most colour alterations have occurred as incidental results of other actions, they have not occurred without commentary. In situations of dramatic change, such as the transformation of the rural to industrial landscapes of the west midlands of England to the 'Black Country' and the green valleys of South Wales to industrial centres, artists, novelists and social historians have commented frequently on the significance and meaning of the change of landscape colour. Only occasionally does one find commentary on changes in other landscapes. In the western rangeland of the United States, Vale deplored the passing of the soft, grey, pleasant smelling sagebrush as it becomes replaced with range management projects of green grass like any other "tamed landscape."[14]

While most colour changes on a broad regional scale have been incidental side effects, the product of individual decisions, there are instances of purposeful colouration. Where governments have instituted regional land use zoning to preserve agricultural or parkland, a conscious effort has been made to preserve "greenness" sometimes using the term 'green belt.' The term and colour has become synonymous with organic, living things, healthy living and aesthetically pleasing environments: the garden city idea.

For the most part, purposeful colouration is a foreground or microscale activity, generally where plants have been selected for flowering or leaf colour qualities. This activity has occurred since prehistoric times. In the nineteenth century, plant hybridization, greenhouses and the introduction of exotic plants made possible even greater massing of bright flowers and shrubs in the parks and gardens of cool monochrome regions. Gradually an attempt to arrange plant colours so as to enhance their visual appeal or taste evolved, particularly through the work of Gertrude Jekyll. The development of gardening to achieve areas of concentrated colours with bright hues balanced with subdued monochromes clearly illustrates the preference in most humans for bright colours in certain situations. Such areas of concentrated bright hues can be designated as *colour points*. Conversely, current standards of taste in green lawns often leads to an obsessive determination to eliminate the discolouring blemish of the lowly dandelion, daisy buttercup, and forget-me-not. Colours are meant for controlled display.

Interest in colours also extends to the built environment where regional tastes and traditions in vernacular architecture indicate preferences for certain wall and roof colours and even restrictions on certain materials or rendering. The red roofs of the Basque region of southern France and northern Spain add much to the colour characteristics of that landscape. In Great Britain in several protected landscapes, all mobile homes (caravans) that are *in situ* have to be green or brown to provide a more harmonious colour context. In recent times there have been fashions in and restrictions on the use of certain building materials and their associated colour. The grey and white granites and limestones have given way to grey and white concretes or tinted concretes. Copper, bronze and coloured glass have been used with colouration in mind. Frequently, climbing and hanging plants are used to soften the stark brightness of modern buildings in a contrived nature-in-captivity context.

It can be seen that aside from the temporal variation in landscape colour of day, month and season there have been and continue to be significant long-term changes in landscape colour which are both incidental and specific to human activity.

THE MEANING OF COLOUR AND LANDSCAPE

Although humans are gifted with colour vision, their appreciation of landscape colour is generally limited. Because of the previous technical difficulties of recording an ephemeral phenomenon, only recently have cheap and widespread colour reproductions of landscape become possible and thus widened the potential for appreciation. Thus, traditional descriptions in word or painting have been approximations for, and by, an elite. Although clearly not the only factor involved, this explains perhaps why the aesthetic appreciation of colour and landscape is limited and rudimentary. Few regional geography texts for example even mention landscape colour let alone give a detailed description of variation and viewing conditions. However, the average citizen sees landscapes in colour and colour reproductions are now widespread. Furthermore, people prefer images in their natural colour.

How might the general appreciation of colour and landscape be developed? As a first step, the differentiation of the variables associated with the three main factors of object, viewing conditions and observer is necessary. Then, together with accurate recording and measurement the generalizations pertaining to geographic variability become feasible such as identifying regional *colour dominants* and *colour points*. A reference system is necessary for a semblance of order to be made in the information available on landscape colour. This must include the factors of perspective, scale and time frame.

The significance of landscape colour is demonstrated by the brief discussion of changes in landscape colour. The value in something is sometimes only appreciated after it has gone. Landscape colour clearly has great value as an indicator of landscape quality yet until recently the technical problems of recording, measurement, reproduction and hence generalization and communication of meaning have not been generally possible. It is now feasible that the general appreciation of landscape colour can be enhanced.

REFERENCES

1. The literature on landscape evaluation from a 20th century objective or rational positivist perspective can be seen in PENNING-ROWSELL, E. "Fluctuating Fortunes in Gauging Landscape Value," *Progress in Geography*, 5, 1981, pp. 25-41. From Giovann Bellini (1430-1516) onwards, one finds the evolution of landscapes treated romantically, 'as states of the soul,' i.e., aesthetically. VALSECCHI, M., *Landscape Painting of the Nineteenth Century*. New York: Graphic Soc, 1969, p. 16. Even earlier, Francesco Petrarch (1304-1374) is reputed to have been the first person to climb a mountain for its view.

2. STANDON, A., (ed.) *Kirk-Othmer, Encyclopedia of Chemical Technology*. (2nd Ed.) Vol. 5. New York: Wiley, 1963, p. 763.

3. HOCHBERG, J., *Perception*. Englewood Cliffs: Prentice Hall, 1964, pp. 20-23.

4. *Ibid.*

5. KALMUS, H., *Diagnosis and Genetics of Genetic Defective Colour Vision*. London: Pergamon, 1965.

6. POST, R.M., "Population Differences in red and green colour vision, deficiency: a review", *Eugenics Quarterly*, 9, 1962, p. 131.

7. Two of the major contributors to the study of *colour* are: CHEVREUL, M.E., *The Principles of Harmony and Contrast of Colours and their Applications to the Arts*. New York: Reinhold (Reprint), 1967. BIRREN, F., *Principles of Colour*. New York: Van Nostrand, 1969. An attempt at understanding the environmental context can be found in MINNAERT, M., *Light and Colour in the Open Air*. London: Bell, 1940.

8. MIDDLETON, W.E., *Vision Through the Atmosphere*. Toronto: U. of Toronto Press, 1958, p. 169.

9. The literature on European landscape painting is extensive — a place to start is CLARK, K., *Landscape Painting*. New York: Scribner, 1950. A focus on colour in landscape art can be found in ITTEN, J., (transl. E.V. Haagen). *The Art of Colour*. Stuttgart: Reinhold, 1961.

10. See BERLIN, B. and KAY, P., *Basic Colour Terms, their Universality and Evolution*. Berkely: U. of California, 1969.

11. BIRREN, F., *op. cit.*, p. 111. See also TUAN, Y-F., *Topophilia*. Englewood Cliffs: Prentice Hall, 1974, pp. 24-27.

12. *Gazetteer of the British Isles*. 9th Edition (Reprinted). Edinburgh: Bartholomew, 1970.

13. *Gazetteer of Canada*. (2nd Edition) Ottawa Surveys and Mapping Branch, Energy, Mines and Resources, 1974.

14. VALE, T.R., "The Sagebrush Landscape," *Landscape*, 2, 1978, pp. 31-37.

15. MASSINGHAM, B., *Miss Jekyll: Portrait of a Great Gardener*. Newton Abbot: David and Charles, 1973.

PLATE 13 Rogers Pass, close to Glacier National Park, British Columbia. ►

6 POSTCARD LANDSCAPES: AN EXPLORATION IN METHODOLOGY

John Marsh

INTRODUCTION

This paper explores the potential use of mass-produced photographs in understanding past, present and evolving public landscape preferences. It focuses on the use of postcards to investigate the past behaviour patterns and landscape preferences of tourists in Glacier National Park, British Columbia. However, as the title indicates, it is intended primarily as an exploration in methodology.

The approach discussed here relates to previous work using photographs to assess landscapes and public landscape preferences but differs in several aspects.[1] Photographs were used by Shafer and associates, for example, "to identify what quantitative variables in photographs of landscapes were significantly related to public preference for those landscapes."[2] Their method was intended for use by natural resource planners required to "make decisions on a factual basis about purchasing, developing or preserving these features." Such photographic methods have been applied and refined but have also been criticised.[3] Lowenthal has gone so far as to suggest that such landscape evaluations are "wholly inappropriate in studying landscape preferences and attachments."[4] He does, however, state that "landscapes that gain public approbation are much reproduced,"[5] that "favoured landscapes and localities are celebrated in literature and painting" and that photography, amongst other things, can "suggest the popularity of this or that generic landscape type, this or that country or city."[6] Accordingly, further experimentation with photographic methods seems hazardous but desirable. It should be emphasized that the method discussed here is not intended for use by natural resource planners; it pertains more to past public preferences for landscapes and relies on the existing availability of mass-produced photographs. To place the method in context, some information on the relationship between photographs and landscape and tourism experiences, and on the evolution of landscape photography will first be provided.

PHOTOGRAPHS, LANDSCAPE AND TOURISM EXPERIENCES

Given that humans are a visually oriented species, it might be hypothesized that landscape photographs would be an important influence on our landscape perceptions and preferences. Thus, Lowenthal argues that we may "feel attracted to landscapes because photos or paintings of them have impressed us"[7] and more specifically, that "countless pictures of canyons and waterfalls influence our ideas of how the Grand Canyon and Niagara Falls ought to look."[8] Sontag both concurs and warns when she says: "photographs create the beautiful and over generations of picture taking — use it up."[9] It might also be suggested that the landscape photographs we take or purchase would reflect our landscape perceptions and preferences. Thus, landscape photographs both influence and reflect landscape perceptions and preferences (Figure 1,6).

If photographs are an important link between man and landscape they might be particularly significant in the links man makes with landscape through tourism. This contention is supported by Sontag who notes that: "photography develops in tandem with one of the most characteristic of modern activities: tourism."[10] The relationship may be examined in more detail with reference to Clawson's five stage model of the outdoor recreation or tourism experience (Figure 2,6).[11] The fact that the sophisticated business of marketing tourism destinations makes great use of photographs in brochures suggests they influence those anticipating a tourism experience. When travelling and at their destination, tourists frequently take or buy photographs. Thus, Sontag notes: "It seems positively unnatural to travel for pleasure without taking a camera along."[12] However, typically, she finds photography, as a link between man and landscape, both a constraint and an opportunity.

> A way of certifying experience, taking photographs is also a way of refusing it — by limiting experience to a search for the photogenic, by converting experience into an image, a souvenir. Travel becomes a strategy for accumulating photographs.[13]

Finally, photographs serve the tourist as a means to recollect landscape and activities, and to relate them to others. Again, to quote Sontag:

> Photographs will offer indisputable evidence that the trip was made, that the program was carried out, that fun was had. Taking photographs fills the same need for the cosmopolitans accumulating photograph trophies of their boat trip up the Nile or their fourteen days in China as it does for lower middle class vacationers taking snapshots of the Eiffel Tower or Niagara Falls.[14]

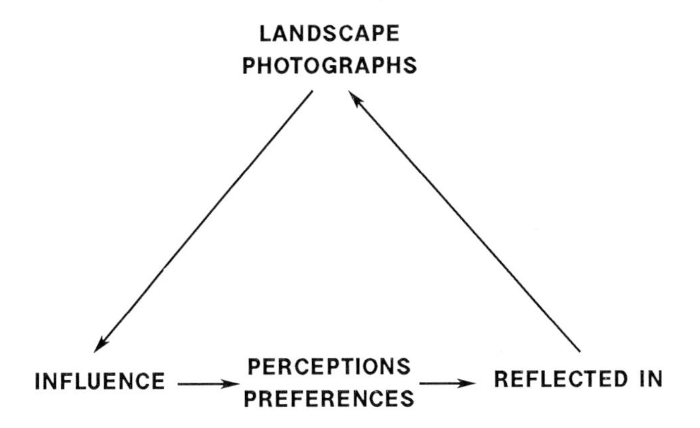

FIGURE 1,6 The Relationship Between Landscape Photographs and Landscape Perception and Preferences

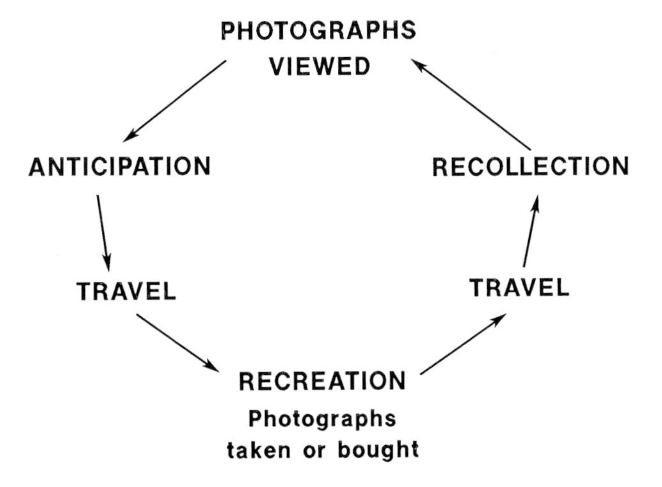

FIGURE 2,6 Photographs and the Outdoor Recreation Experience

EVOLUTION OF LANDSCAPE PHOTOGRAPHY

Photography developed in the mid-nineteenth century from the daguerreo-type process, invented in 1829, which produced a single image on a copper plate, to the wet-plate, collodion, process of the 1850s, which produced negatives on glass that could then be used to produce multiple paper photographs. Despite such progress, the cameras and processing techniques were still so cumbersome and crude that all but the most dedicated professional photographers were deterred, especially away from urban centres, from indulging in photography. However, as Birrell notes:

> by 1880 the gelatine dry plate had revolutionized photography. The great ease in using the various dry processes lead to increased use of the camera on Canadian surveys ... (and) the transparent celluloid base roll film developed by George Eastman in the late 1880's finally made the camera a household item.[15]

Given such improvements, landscape photographs could be used, amongst other things, to examine potential routes for the Canadian Pacific Railway, (CPR) to survey the Rocky Mountains, to depict areas for settlement and to encourage tourism. The age of photographic influence on public awareness and appreciation of landscapes had begun.

Photographs were used to advertise the tourist attractions along the CPR, especially in the Rocky and Selkirk Mountains. For example, the firm of William McFarlane Notman of Montreal was one of several encouraged or contracted by the CPR to take photographs of the scenery along the route.[16] Notman's photographs were used in preparing the British Columbia segment of that popular work, *Picturesque Canada* and Hart notes "these had been valuable in providing the Canadian public and potential tourists, with one of their first views of the country that the CPR was about to open up."[17] When the line opened in 1886 more photographers, such as Buell and Henderson, travelled the route taking photographs for public slide shows, magazine and book illustrations and tourism advertising. Soon folios of photographs were being sold as souvenirs to the burgeoning number of tourists.[18] By 1900, the mass production of landscape photographs for popular consumption, especially by tourists in scenic areas such as the Rockies, was not only feasible, but commonplace.

One such type of mass-produced photograph is the postcard. The postcard evolved and became popular partly as a function of improved photography and the growth in tourism and partly as a result of post office initiatives and changing regulations. The forerunner of today's postcard the postal stationary card was first produced in Canada in 1871.[19] Sold by the post office, the cards were to have a stamp and address on one side and a message on the other, thus leaving little room for illustration. In the 1890's privately produced postcards became legal and some illustration became legal and common on either the address or message side of the card. In 1903, to quote Steinhart, "one of the most important changes in the postal history of the postcard in Canada occurred."[20] Divided back postcards, having the address and message on one side became legal, hence, the complete reverse side was available for illustration. As a result, as Anderson and Tomlinson note, "manufacturers, who had already honed their skills to a fine point with coloured printed material of other kinds, turned to the production of postcards with vigour."[21] Postcards became enormously popular and numerous companies both in Canada and abroad became involved in their production. Valentine and Sons of Scotland, alone, turned out 20,000 Canadian views. In 1900, Canadians mailed over 27,000 cards, by 1908, 41,000,000 were mailed, and by 1913, over 60,000,000.[22] Steinhart[23] refers to the Edwardian period as the "golden age" of the postcard, but while the First World War changed many social habits postcards have remained a popular artifact in our society.

Old postcards have survived and are collected, with postcard clubs facilitating preservation and interchange of cards and research on them.[24] Accordingly, there are numerous large photographic data sets available for researchers interested in using photographs to evaluate landscape or to investigate landscape preferences. To study such data it may often be necessary to sample the data set or focus on a particular theme. Attention might be directed, for example, at views of one particular place or landscape type, at views produced by one photographer or publisher, or at views taken over a particular time period. The author chose to look at 140 postcards of one particular place, Glacier National Park, produced over a period of 22 years by a variety of publishers.

LANDSCAPE PHOTOGRAPHS OF GLACIER NATIONAL PARK

The Glacier National Park area of British Columbia is a scenic mountain area along the Canadian Pacific Railway. An area around Rogers Pass, Mt. Sir Donald and the Great Glacier was designated a national park in 1886, Glacier House Hotel was built, and tourists from all over the world began passing through or staying in the park (Figure 3,6). A few photographers, such as Notman, arrived before the line was open, but many more came afterwards and soon the area was depicted in brochures, books and postcards.

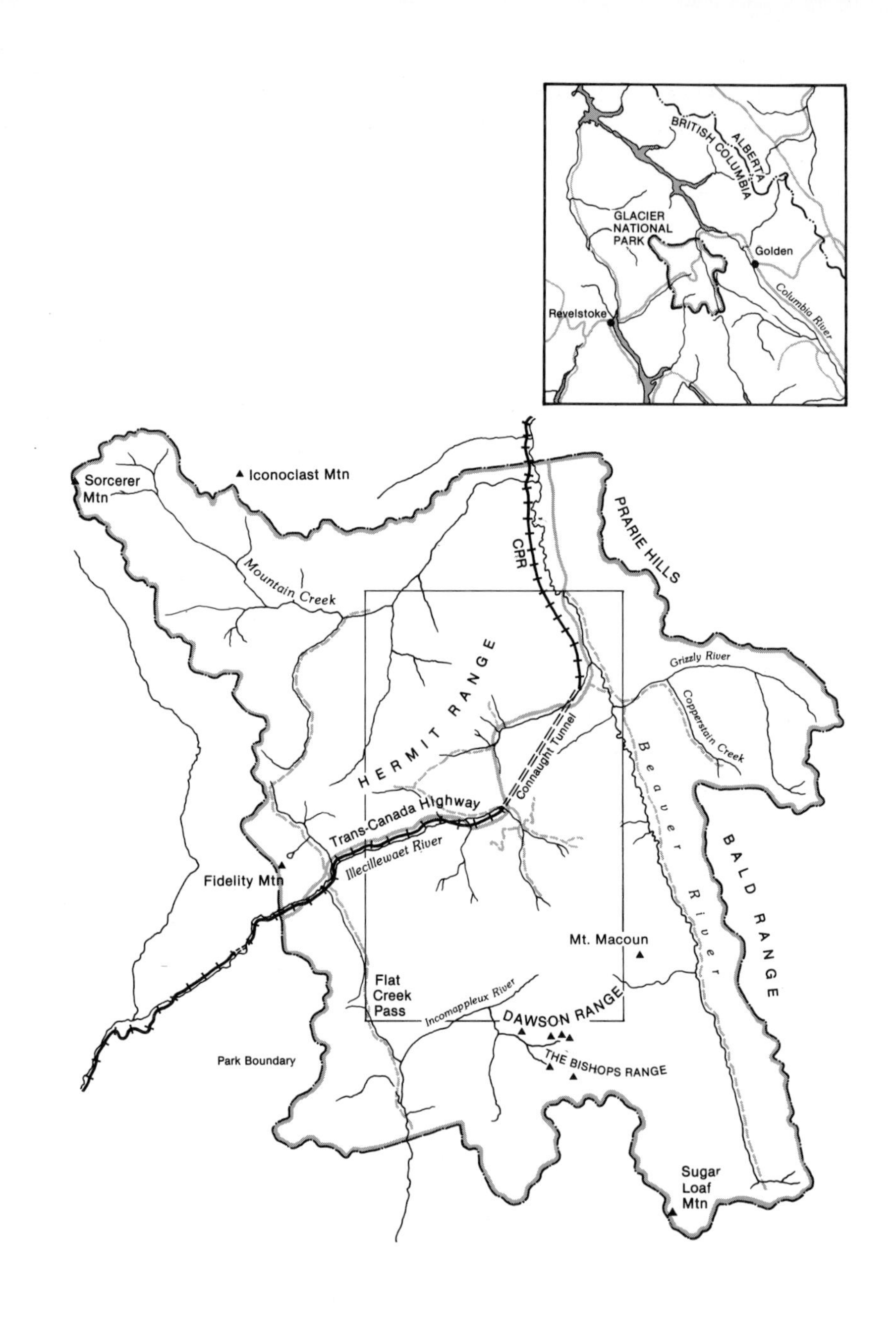

FIGURE 3,6 Glacier National Park: Area Photographed 1903-1925

In the process of conducting research on the evolution of the park, the author collected historical photographs, primarily postcards, that could provide evidence of landscape and tourism characteristics.[25] More recently, the author has endeavored to collect one copy of every view produced so as to obtain a representative photographic data set for the period 1903 to 1925. The year 1903 was selected because it was the year the picture postcard was legalized; 1925 was selected, as the last year the Glacier House Hotel operated. Thereafter, until the 1960s, there was negligible tourism in the park. Over a period of fifteen years, 140 postcards of the present park area produced by publishers and dating from this period have been collected and these form the data set for the analysis described below.

Photographic Data Analysis

The photographic data was analysed in five ways. First, spatial information was abstracted to answer such questions as: what area of Glacier National Park is included in the view; what areas were photographed most; and from what locations were photographs taken? Second, content analysis was undertaken to answer such questions as: what specific natural features were photographed; what specific cultural features were photographed; what was the ratio of views of natural features, to views of cultural features, to views of both; what proportion of photographs included people and what activities were shown? Third, temporal information was obtained to answer such questions as: in what season were photographs taken? Fourth, technical information was procured to answer such questions as: what was the ratio of horizontal to vertical views; and what proportions of the data set were black and white, a single colour, or multicoloured? Fifth, the photographs were analysed to determine which areas of the park, specific natural or cultural features, and activities were *not* depicted. In addition, the proportion of cards with messages on and those without was noted, and an analysis made of the messages themselves. The analysis was thus in part quantitative, and in part qualitative, it being possible to answer some questions precisely, others with less precision but consistently, and still others with great difficulty. The problems encountered will be dealt with in a discussion later of overall methodological weaknesses and difficulties.

Results

The results of the analysis are summarized in Table 1,6 and Figure 4,6, and will be described with reference to the above questions.

Table 1,6 RESULTS OF POSTCARD ANALYSIS OF GLACIER NATIONAL PARK

Spatial Information

1. Eight percent of park area is included in views.
2. The most frequently photographed areas are the railway corridor between Stoney Creek and Cougar Mountain, the upper Illecillewaet Valley, Asulkan Valley, and surrounding peaks.
3. Photographers' location: Mt. Abbott trail 45, Great Glacier trail 21, Avalanche Crest trail 9, railway near hotel 7, the 'Loops' 5, Stoney Creek 5

Content Information

4. Specific natural features photographed: Mt Sir Donald 26, Great Glacier 21, Loop-Illecillewaet 18
5. Specific cultural features photographed: Glacier Hotel/Mt Abbott 14, Stoney Creek Bridge 5
6. Of total 140 cards, 12 included people (5 on glacier, 3 at glacier snout, 3 on horseback, 1 climbing)

Temporal Information

7. Seasonal — only 1 winter view
8. Daily — mostly high sun conditions

Technical Information

9. Ratio of horizontal to vertical views 78:62
10. Black and white 34, sepia 7, blue 1, multicoloured 98

The photographs depict approximately eight percent of the present park area of 1349 square kilometres. The coverage is almost exclusively along the railway corridor, more specifically that section from Stoney Creek, southwest for 16 kilometres to Cougar Mountain (Figure 4,6). The only area away from the railway corridor to be photographed was that containing the headwaters of the Illecillewaet River, the Asulkan Valley and the surrounding mountains. The area photographed was accessible by train or rail. However, other areas, such as the Illecillewaet Valley west of Cougar Mountain and the lower Beaver Valley, that were also accessible by train or trail were not photographed. Particularly popular as locations from which to take photographs for postcards were: the trail from the hotel up Mt. Abbott, the trail to the Great Glacier, and the trail up Avalanche Crest. Four more specific locations were particularly favoured: a point on the Mt. Abbott trail giving a view of the Great Glacier and Mt. Sir Donald, a view-

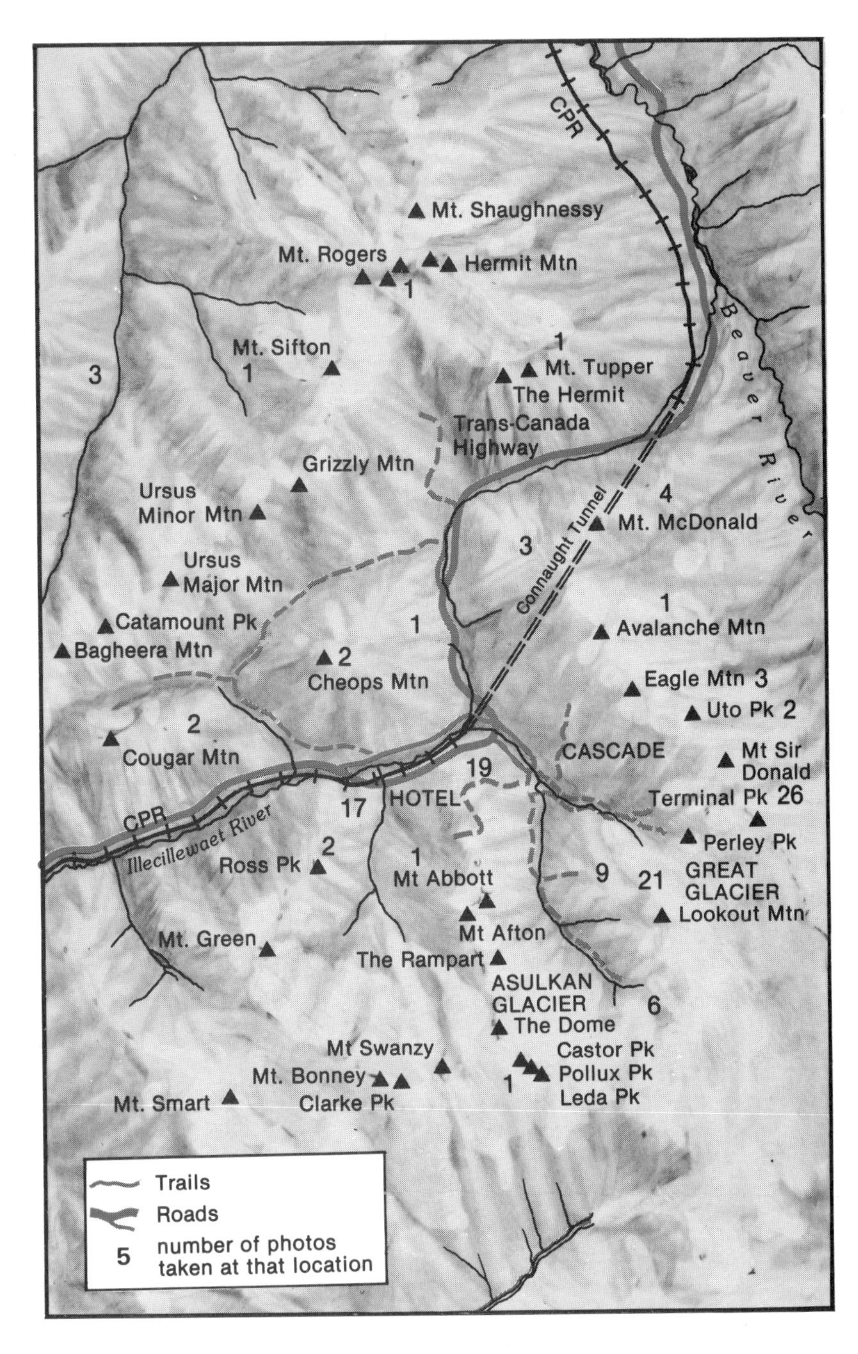

FIGURE 4,6 Glacier National Park: Features Photographed

point at Marion Lake overlooking the Illecillewaet Valley and Rogers Pass, a viewpoint on the Great Glacier trail showing the Illecillewaet River, boulders and the glacier nearby, and a viewpoint at the top of the waterfall on Avalanche Crest from where the hotel and Mt. Abbott above could be photographed. All such locations were readily accessible by trails from the hotel in under two hours.

In terms of content, the most frequently depicted natural features were Mt. Sir Donald, the Great Glacier and the Illecillewaet Valley and Cougar Mountain (Table 1,6). Mt. Sir Donald, the highest peak in the area, likened to the Matterhorn and regarded as a great challenge to mountaineers, was clearly visible from the Glacier House Hotel and station and from the railway approaches via Rogers Pass and the Illecillewaet Valley (Figure 5,6). The Great Glacier, also visible from the hotel and station, was the most accessible Glacier along the route of the railway through the mountains, and could be reached on foot or horseback in less than an hour.

The single most frequently photographed cultural feature was the Glacier House Hotel (Table 1,6) which was sometimes depicted close-up from the railway, or from above with Mt. Abbott rising above it, or from down valley with the Great Glacier or mountains in the background (Figure 6,6).[26] The second most photographed cultural feature was the Stoney Creek Bridge, located on the west side of the Beaver Valley. This bridge, one of the highest on the line, was initially a wooden trestle rather than a steel arch, and was usually photographed from below, presumably for dramatic effect. Given that trains normally did not stop here, and it was a considerable hike from the hotel, few tourists would have inspected it from any angle. Next in frequency of cultural features depicted came the trestle bridges known as the "Loops" because of their configuration, and then the entrances to the Connaught Tunnel, which opened in 1916 and eliminated a hazardous section of track over Rogers Pass and past the Hotel.[27] An individual card of unusual interest is one depicting the setting of the Connaught Tunnel, its east portal and a map showing its location erroneously (Figure 7,6).

An attempt was made to classify views as natural, cultural or both. This resulted in 63 views of natural features, 3 views of cultural features and 74 views having both features. Common amongst views having both features were ones including the railway and mountains and valleys and ones of the hotel with a mountain or glacier background. Few pictures of cultural artifacts were taken without their natural setting being included.

Relatively few photographs included people, there being only 12 out of 140 (Table 1,6). Of these, 5 showed tourists on the Great or Asulkan Glaciers, 3 depicted tourists, often with horses, at the snout of the Great Glacier, 3 showed tourists on horseback travelling one of the trails, and one depicted a group rock climbing (Figure 8,6).

FIGURE 5,6 Reproductions of Postcards of Natural Features

FIGURE 5,6 (Continued)

FIGURE 5,6 (Continued)

FIGURE 6,6 Reproductions of Postcards of Cultural Features

FIGURE 6,6 (Continued)

FIGURE 7,6 Reproductions of Postcards of Railway Features

FIGURE 7,6 (Continued)

FIGURE 7,6 (Continued)

FIGURE 7,6 (Continued)

FIGURE 8,6 Reproductions of Postcards of People and Activities

FIGURE 8,6 (Continued)

FIGURE 8,6 (Continued)

Information was also obtained on when photographs were taken. Most appeared to have been taken in summer, there being little evidence of the snow that persists at low altitudes until June or of the fall colours of deciduous trees in the area. The hotel was normally closed during the late fall, winter and early spring and few tourists or photographers appear to have passed through the park at this time. Most photographs appear to have been taken during the middle of the day, long shadows being absent. A small number of cards also appear to have been artistically enhanced, showing scenes, such as Mt. Sir Donald, by moonlight.

Given the extremes of topographic relief in the park, it might have been expected that vertical views would have been more common than horizontal ones. In fact, the opposite was the case; 78 of the 140 views were horizontal. Despite the infancy of photography at the turn of the century, most of the cards (98) were multicoloured, a few were sepia or blue and the rest (34) black and white. There was great variability in the accuracy of colour renditions and evidence of artistic enhancement, such as the green tinting of rocky and burnt over areas, and the red tinting of hotel roofs.

Something can be learned by considering what areas and features of the park, and which tourist activities, were *not* depicted in the data set. Numerous specific features could be mentioned but only some of the more obvious omissions will be noted. In the railway corridor neither the lower Illecillewaet Valley or Beaver Valley were featured, and outside of the railway corridor, with the exception of the hotel area, nothing was photographed for postcards. Access to the backcountry areas was very difficult and time-consuming, and few tourists except intrepid mountaineers went there. Surprisingly, few photographs were taken of the Cougar Valley and Nakimu Caves area which was a relatively popular tourist destination, and accessible in later years by carriage. Despite the overall frequency of photographs including railway features, there are none of the large bridges over Mountain and Surprise Creeks in the Beaver Valley and very few of trains. There are no photographs of wildlife, either birds or animals, though plenty existed in the area. Although there are some photographs of people there are none in this data set of tourists at the hotel or station, none of people picnicking, caving or skiing, and none of people inside buildings. Finally, while in a few cards the sky is tinted pink in places, there are no real sunset pictures.

When the author began collecting these cards, it was hoped that the messages on the reverse might be as informative regarding tourist appreciation of the park landscape as the photographs themselves. However, only 45 of the 140 cards had messages on them, and only 40 of these cards had actually been mailed as postcards. Nevertheless, a few comments on the messages are warranted. The messages indicate that of the 40 cards mailed some were sent before reaching the park and others long after leaving it. Only 23 cards contained comments on the park scenery, however, several seem particularly relevant to the subject of this paper:

> The beauties of the scenery at Glacier cannot be photographed. It is sublime. I was awed into silence (which seldom occurs) at the grandeur of the scene.
>
> This is typical of the scenery we have had going through the Canadian Rockies.
>
> Arrived here this morning, this is a great trip. Just like Switzerland.
>
> Oh such wonders. Our window looks out on this view (Mt. Sir Donald and Great Glacier). The mountain streams are beautiful. The caves are wonderful.
>
> Well I am out in the wilderness at last. Nothing but mountains on both sides.
>
> We have taken this out (the Loop) and put in a steel bridge.

Interpretation

As previous users of photographic analysis have found, it is one thing to categorize and quantify the photographs and their content, but another to interpret the meaning of the results so as to be confident of an accurate understanding of landscape appreciation.

For the moment, let me presume that the data set and the results of analysis should be indicative of tourist behaviour and landscape appreciation. If so, the results suggest a number of conclusions.

In terms of behaviour, tourists seem to have concentrated in the railway corridor, which for most was the means of access to the park. In particular, tourist activity was focussed on the Glacier House Hotel and on the short trails radiating from it, particularly to Marion Lake, the Great Glacier and Avalanche Crest. Use of the backcountry was negligible but a few tourists visited the Nakimu Caves. Tourists, both men and women, came predominantly in summer and undertook activities such as glacier exploration, horseback riding and climbing.

In terms of landscape preferences, tourists were impressed by three general landscape features: mountains, glaciers and valleys. More specifically, they were impressed by Mt. Sir Donald, the Great Glacier and the Illecillewaet Valley. While landscapes lacking human artifacts were deemed appealing, there was considerable appreciation of landscapes blending man and nature, such as the setting of the hotel and the section of the Illecillewaet Valley with its looping railway trestle bridges. Both the hotel and the Stoney Creek bridge were features of special interest. Several views were regarded as especially beautiful by moonlight.

The evidence of logging, forest fires, and the railway architecture and operations depicted in the photographs, do not appear to have detracted from tourist appreciation of the landscape. In contrast, Lowenthal noted perceptions in England of "the evil effect of railroads and engineering architecture," which effect, according to Hamerton writing in 1885, operated "rather by its suggestion of hurry and business to the mind than by real offence to the eye."[28]

These are some preliminary, rather general interpretations of the results. One means of checking the validity of such interpretations would be to compare them with information on tourists' behaviour and landscape preferences obtained by other methods such as literary analysis and investigation of tourism agency records. This type of analysis has already been undertaken by the author[29] and such a comparison is possible for this place and time period. However, a discussion of such a comparison, and the evidence it provides for testing the validity of the interpretations of tourist behaviour and landscape preferences deduced by this photographic analysis are beyond the scope of this paper. Suffice it to say, that many of the interpretations regarding tourist behaviour and landscape preferences in Glacier National Park between 1903 and 1925 reached by this photographic method accord with those reached by the literary method. There are, however, differences and one might will question the validity of both methods.

METHODOLOGICAL PROBLEMS

As this chapter is intended primarily as a methodological exploration it is important to consider the conceptual and practical difficulties it poses.

First, as with any quantitative technique, there is the problem of judging the adequacy of the data set and the representativeness of the sample available or selected. In this case, a sample of 140 photographs may seem quite considerable, but it is conceivable that thousands of different cards depicting this area were produced over this period. This might be assessed for various publishers by analysing the photograph numbers that appear on most cards. Cards of one area were often numbered sequentially, though not always, hence it might be possible in some cases to determine the total number produced and from that determine the proportion in the sample obtained. Occasionally, publisher's catalogues have survived, or postcard collectors have compiled them, thus making assessment of the sample easier. Another approach to check on the sample would be to compare it with other collections for the same place and period.

In obtaining a sample by collecting there is always the possibility that postcards produced in limited numbers will not have survived or will rarely be found on the market. Only diligent collecting of all postcards found over a long period can help overcome this problem.

Second, is the problem of knowing whether all the postcards published were actually bought. Both bought and unbought postcards may have survived and be indistinguishable, as not all bought cards were written upon or mailed. If we are to try and judge landscape preferences from the post cards bought, then this clearly is a difficulty. While catalogues of historic cards are scarce, sales records for particular cards are nonexistent.

Third, is the problem of knowing who bought the postcards. While post cards were a popular artifact in the period studied here, presumably not everyone bought them. They may be more reflective of public taste than the much more limited number of literary accounts, yet still it is unclear just what proportion of the tourist population's views are indicated by this analysis. A study of current postcard buying patterns may throw some light on this problem.

Fourth, is the problem of knowing why people bought which postcards. Judging from a common understanding of present buying patterns it seems some people may have bought cards of places they had yet to visit or would never visit and of views they had not actually seen, and of activities they would never undertake. Perhaps, more importantly, people may have bought the postcards that impressed them as postcards, rather than as representations of the landscapes before them. Thus, Lowenthal notes: "most photographs are *meant* to be pictorial, and scenes deliberately posed and framed are inevitably viewed more as pictures — that is, works of art — than in the myriad ways we regard landscapes."[30] He goes on to say: "Landscape judgements based on surrogates may conform with choices made on site, but the concurrence does not remove the distinctions." It is quite conceivable therefore, that people bought postcards that impressed them as artistic representations, that conformed to their views of ideal landscapes or activities, or that would serve as impressive souvenirs for themselves or others. As such, postcards may tell us something about public landscape preferences but less about the preferences for the landscape being studied.

Fifth, is the problem of knowing how influential the photographers were in depicting the landscape as a beautiful place and in deciding the range of postcard views, representative or otherwise, that were to be available for public consumption. Do the postcards say more about the landscape preferences of the photographers than the public? Presumably, landscape photographers were often influenced by photographic fashions, thus Lowenthal observes that: "landscape photography has reflected several ideals, emphasizing in turn prettiness, verisimilitude, contrasts of light and texture, collage and superimposition, movement and transience."[31] It has also reflected the availability of various types of equipment and processing methods. The lack of postcards depicting wildlife might be attributed, therefore, to public taste or photographic fashions or the unavailability of high speed cameras and films. Perhaps, these difficulties can be partly overcome by using the postcards of a wide range of photographers, and by recognizing that while the photographers limited the choice of views and artistically enhanced them, nevertheless, considerable opportunity for public choice, based on landscape preferences remained, including the option of not buying them.

Sixth, is the problem of analysing the photographs in the data set. In the course of attempting a quantified analysis of the photographs used here a number of difficulties were encountered. It is difficult to date postcards particularly those without postal dates indicated. This could sometimes be overcome by considering the style of card and sometimes with reference to landscape features, such as sections of the hotel and the railway tunnel that were datable. A problem exists also in deciding how many postcards could be considered original and distinctive and how many were reprints or enhancements, and therefore, in deciding upon the total sample to be used. Whenever a postcard exhibited a definite difference from another, say in the content of the view, the caption or "addition" of moonlight, it was included in the sample. In analysing the content, there is considerable difficulty in categorizing cards as showing natural or cultural features or both. When different people were asked to undertake this categorization different results were obtained, though the discrepancy was less than ten percent. Clearly more precise guidelines for undertaking this categorization are needed if the results of categorization are to be closely replicable and discussion of the results is to be meaningful.

Finally, there is the overall conceptual difficulty of knowing whether, given all the other problems and their potentially synergistic effect, the production, quantity and content of views really are indicative of tourist behaviour and landscape preferences. Only further consideration of this overall difficulty, cross-checks with other verifiable sources of information, and studies of present postcard buying patterns and their relationship to tourist behaviour and landscape preferences can help resolve this.

FUTURE RESEARCH

A number of research questions and research themes are suggested by the methodological exploration discussed here. First, research is needed to assess the availability of similar data sources and to examine the problem of obtaining, and knowing one has obtained, a representative sample. Second, further consideration should be given to the means of analysing such data sets, the variables to be considered, and the means of categorization. Third, studies should be undertaken to determine who bought post-card, who buys them today, and why, and how buying patterns relate to tourist behaviour and landscape appreciation. Information might be sought from publishers and retailers regarding how cards are selected, how distributed and which sell most effectively. Fourth, research might be undertaken on trends in the production of postcards in their content and in the public's use of postcards. This might contribute to increased under-standing of changes in tourism behaviour, in preferences for places, and in landscape preferences through time. Fifth, research on current postcards and the patterns of consumption might be related to that undertaken using other methods for evaluating landscapes and public landscape preferences. Sixth, this research could be extended to look at other past and present photographic data sets such as calendars, coffe table books, posters, adver-tisements and private photographic collections and family albums.

In summary, the author agrees with Lowenthal that: "mass preferences can be deduced only in part from the popular appeal of surviving land-scapes and of books and pictures that celebreate their charms."[32] How-ever, landscape photographs, as Sontag says[33] are "pellets of information" that when available in quantity as are postcards, could be analysed, hopefully, to yield some useful and comparable information relating to past or present public behaviour and landscape appreciation. At least further in-vestigation and discussion of this data base and methods for its analysis seems warranted.

REFERENCES

1. For example see early papers by PETERSON, G. and NEUMANN, E., "Modelling and predicting human response to the visual recreation environment", *Journal of Leisure Research*, 1, 1969, pp. 219-237 and SHAFER, E.L., HAMILTON, J.F. and SCHMIDT, E., "Natural landscape preferences: a predictive model", *Journal of Leisure Research*, 1, 1969, pp. 1-19.

2. SHAFER, E.L. *et al.*, *op. cit.*, p. 1

3. For example see KREIMER, A., "Environmental preferences: a critical analysis of some research methodologies", *Journal of Leisure Research*, 9, 1977, pp. 88-97.

4. LOWENTHAL, D., *Finding Valued Landscapes*. Environmental Perception Research Working Paper # 4, University College, London, 1979, p. 24.

5. *Ibid.*, p. 11.

6. *Ibid.*, p. 2.

7. *Ibid.*, p. 54.

8. *Ibid.*, p. 52.

9. SONTAG, S., *On Photography*. New York: Dell, 1973.

10. *Ibid.*, p. 9.

11. CLAWSON, M. and KNETSCH, J.L., *Economics of Outdoor Recreation*. Baltimore: Johns Hopkins, 1966.

12. SONTAG, S., *op. cit.*, p. 9.

13. *Ibid.*

14. *Ibid.*

15. BIRREL, A.J., "Classic survey photos of the early west", *Canadian Geographical Journal*, 9, 1975, pp. 12-19.

16. HARPER, J.R. and TRIGGS, S. (eds.), *Portrait of a Period: A Collection of Notman Photographs, 1856 to 1915*. Montreal: McGill University Press, 1967.

17. HART, E.F., *The Selling of Canada: The CPR and the Beginnings of Canadian Tourism*. Banff: Altitude Publishing, 1983.

18. *Ibid.*, p. 39.

19. STEINHART, A.L., *The Postal History of the Postcard in Canada, 1871 - 1911*. Toronto: Mission Press, 1979.

20. *Ibid.*, p. 40.

21. ANDERSON, A. and TOMLINSON, B., *Greetings from Canada: An Album of Unique Canadian Postcards from the Edwardian Era, 1900 - 1916*. Toronto: Macmillan, 1978.

22. *Ibid.*, p. xiii.

23. STEINHART, A.L., *op. cit.*, p. 61.

24. TORONTO POSTCARD CLUB, *Card Talk*, 1, 1984.

25. MARSH, J.S., "Man, landscape and recreation in Glacier National Park, British Columbia, 1880 to present", unpublished Ph.D. thesis, University of Calgary, 1971.

26. MARSH, J.S., *A History of Glacier House and Nakimu Caves*. Peterborough: Canadian Recreation Services, 1979.

27. WOODS, J. and MARSH, J.S., *Snow War: An Illustrated History of Rogers Pass, Glacier National Park, B.C.* Toronto: National and Provincial Parks Association of Canada, 1983.

28. LOWENTHAL, D., *op. cit.*, p. 40.

29. MARSH, J.S., *op. cit.*, 1971.

30. LOWENTHAL, D., *op. cit.*, p. 54.

31. *Ibid.*, p. 53.

32. *Ibid.*, p. 34.

33. SONTAG, S., *op. cit.*, p. 69.

PLATE 14 A licence plate from Ontario proclaims "Keep it Beautiful", but how much effort is put into this as opposed to keeping it wealthy? ►

PLATE 15 A prairie scene in Saskatchewan. ►

ONTARIO
JKV 081
73 KEEP IT BEAUTIFUL
ONT 80

7 THE USE OF PERSONAL CONSTRUCT THEORY IN EVALUATING PERCEPTIONS OF LANDSCAPE AESTHETICS

J.W. Pomeroy, J.E. FitzGibbon and M.B. Green

INTRODUCTION

Landscape planning has been hampered by the lack of a suitable theory by which the aesthetics of a landscape can be assessed. Presently, the art of assessing landscapes is based on elusive qualities of the landscape, defined by the experience of practitioners and strongly ascribed to by those who have accepted the rules and procedures of the "profession". Landscape architects and planners have devised criteria for the "aesthetic" landscape without reference to scientific measurements of such.

It has recently become more generally accepted among landscape planners and scientists that landscape beauty should be defined by the opinions of the population that uses the landscape, rather than the opinions of an elite profession. Landscapes are "goods" of mass consumption for which the public has no choice but to consume. Their designs should reflect the ability of the public to discriminate among landscapes and the preferences of the public for landscape. Without such a basis, man's designed environments are in danger of being unappreciated and futile, with qualities that only a self-designated elite group can appreciate.

This study proposes personal construct theory, developed in environmental psychology, as the basis for new methodologies in interpreting the perceived similarities among and preferences for various landscapes. These qualities of a landscape can designate unique places, attractive landscapes, ugly landscapes, monotonous landscapes, and areas of great landscape variety. Precise determination of landscape similarity and preference can provide valuable input for planning decisions and the procedures of the professions involved in the difficult task of designing and managing landscapes.

USE OF PERSONAL CONSTRUCT THEORY IN LANDSCAPE ASSESSMENTS

Personal Construct Theory

Kelly[1] developed personal construct theory for use in psychoanalysis of human responses to various stimuli. He proposed that a person's present perception of his or her environment is based upon the complex interaction of past experience in the memory. New experiences are evaluated or "construed" using the order made out of previous experience as a basis. Personal constructs are the criteria used by a person to describe the conceptual structure which is derived from past experience and to interpret new experiences in terms of the existing conceptual structures. Constructs are bipolar concepts which categorize the perceived similarities and differences among environmental stimuli.

The Repertory Grid

Personal constructs of environmental stimuli are most often elicited from individuals using the repertory grid technique.[2] The repertory grid is a matrix showing the disparities among aspects of the environmental stimuli. In the repertory grid technique, each individual discriminates among environmental stimuli. This construing process is based upon the individual's personal constructs of his or her environment. The discrimination process can be achieved by the sorting, ranking or rating of environmental stimuli.[3] Ratings of similarity or preference are compiled in the matrix. The repertory grid for an individual can be combined with the grids from other individuals to form an aggregated "supergrid".[4] The supergrid represents the aggregated personal constructs of several subjects. Those constructs that are consistently important among the individuals will have the greatest influence on the composition of the supergrid.

The repertory grid technique has been successfully applied in measurement of the perception of similarities among, and preference for, landscape paintings,[5] urban sketches,[6] environmental cognition,[7] recreation areas,[8] and landscape photographs.[9] The results of the various studies show that the perceived similarities of landscape representations are based on strong, common personal constructs in cases when the study participants are familiar with landscape. However, there tends to be more disagreement on constructs describing preference for landscape representations.[10]

Cognitive Set

O'Hare[11] found strong differences in preference for landscape paintings between persons trained and untrained in art appreciation. This suggests that "groups" of people with similar personal constructs of landscape preference may exist. Ward and Russel[12] refer to the cognitive set as defining the mental universe of concepts which is represented by personal constructs. The cognitive set is the conceptual structure in a person's mind within which environmental stimuli are assessed. Ward and Russel state that a particular landscape can be assessed in terms of the regional landscape by aligning the cognitive set of the study subjects to the regional landscape. This imitates the mental processes of landscape assessment normally performed by a populace on their local landscape. For instance, a resident of the Prairie Provinces does not commonly assess the "aesthetics" of a wheat field by comparing it to the coast of Vancouver Island or Montreal's city centre; these views are not directly comparable. Landscape scenery tends to be assessed in terms of similar landscapes recently viewed.

The cognitive set can be an extremely useful concept in evaluating landscapes within a planning region because;

1) groups of persons with common cognitive sets in terms of similarity perception and perhaps preference can be found and constructs important in their landscape assessments determined;
2) assessment of a series of landscapes in a planning region familiar to the study participants can often be described by an aggregate cognitive set of the simple personal constructs common to most subjects.

The difference O'Hare[13] has noted in the preferences of art students and others towards landscape paintings can be explained by the altered cognitive set of students studying art appreciation. Similar differences may exist between landscape architects and the populace they serve.

Multidimensional Scaling

Multidimensional scaling (MDS)[14] of repertory supergrids has received increasing support as a suitable method of determining the aggregated personal constructs that compose the repertory supergrid and determine landscape perception.[15] MDS takes the disparities among elements of the supergrid and uses them to create an n-dimensional space. The configuration of this space is such that distances among the elements are as recorded in the repertory supergrid. The dimensions of the n-dimensional space correspond to the aggregate personal constructs used in the differentiation of environmental stimuli.[16] The area of the space occupied by the landscape elements corresponds to the cognitive set common to the subjects.[17]

Use of Surrogate Landscapes

The use of surrogates to represent landscapes has received some criticism[18] but continues to be the most feasible method for eliciting a large number of responses to various landscapes. Photographs have been used and recommended by most researchers examining public response to the aesthetics of landscapes.[19] Shuttleworth[20] provided a comprehensive review and severe test of the use of photographs as surrogates in landscape perception studies. He concludes that there is a sound basis for using photographs in landscape studies provided two criteria are met by the photographs:

1) they must be colour photographs;
2) they must provide the lateral and foreground context in each of the views without distorting the view.

Dunn[21] notes that strict controls on photographic quality and the representatives of composition must be maintained as well.

METHODOLOGIES

The analysis of perception of similarities among, and preference for, landscapes in this study requires methodologies tailored to the information desired from the analysis and the resultant configurations of the MDS space. The methodologies used to elicit a response to landscape photographs and interpret the results of the MDS differ somewhat between the similarity and preference analyses. The adaptability of personal construct theory, the repertory grid and MDS to the two methodologies while maintaining theoretical validity demonstrates a strength of this approach.

Perceived Similarity Data

The similarity analysis is designed to obtain judgments of the perceived similarity of landscapes in a planning region displaying both urban and rural scenes. Common personal constructs derived from the MDS are to be associated with permanent attributes of the landscape.

STUDY AREA AND PHOTOGRAPHIC REPRESENTATION

The "riverscape" of the South Saskatchewan River in, and adjacent to, Saskatoon, Saskatchewan was selected for this study of landscape perception. Saskatoon is the major centre of northern settled Saskatchewan and had a population of 165,000 in 1983. The riverscape bisects the city and includes commercial, industrial, residential, recreational and agricultural land uses. Local terrain is flat to moderately rolling except for the river valley, which is a broad floodplain in the south of the study area, narrowing to a sharp incised valley in the north. The river valley displays water, forest, topographically diverse terrain and the Saskatoon city centre in a region of sparsely populated, typically monotonous, semi-arid prairie.

Forty 35 millimetre colour photographs of the South Saskatchewan River Valley were taken during periods of high sun, on clear, cloudless days in late summer. These conditions characterize those from late May to mid-September when vegetation is flourishing and can be fully assessed as an attribute of the landscape. The photographs show foreground and lateral context as well as the unobstructed background landscape. The photographs were taken from accessible, well-travelled viewpoints along bridges, riverbank drives and trails. Composition includes both banks looking along the river, with the sky to land ratio constant at 1:5.

TESTING PROCEDURE

The participation of thirty University of Saskatchewan students was enlisted using a campus-wide advertisement. The students are from a broad range of disciplines and backgrounds, of ages 19-45 and proportionately representative of the sexes. This group of people may be representative of the young to middle-age population of Saskatoon with a Grade 12 or greater education. This population includes the most intensive users of the Saskatoon riverbanks.[22]

The participants were given a randomly mixed stack of the 40 nine by thirteen centimetre colour prints of the river valley landscape and informed this was a "landscape study". They were instructed to sort the photographs into as many piles as they wished based on any criteria. The piles produced are similarity groupings of the photographs; similarity derived from unbiased evaluation of the landscapes. The number of piles and the photographs placed in each pile were recorded by the experiment supervisor.

LANDSCAPE ATTRIBUTES

The photographs were visually inspected for selected attributes: colour, angle of view, vegetation, clarity, land use, valley slope, soil exposure, urban blight and attempts at "enhanced" urban architecture. These attributes were selected from the range of "aesthetic factors" postulated by Leopold,[23] Linton[24] and Litton *et al.*,[25] though selection was based on the prominence of the attributes in the set of landscapes studied.

The landscape attributes were determined qualitatively for each photograph. Ranges of the attributes within the set of landscapes were observed. Labels of positive and negative were assigned to the extreme ranges of the bipolar gradations. The non-determinant label was assigned to the centre of the gradations. Each photograph was assigned a "plus", "minus" or "zero" for each attribute. A photograph received a plus or minus if it strongly possessed a range of an attribute and zero if it did not or the attribute did not apply to the scene.

While this method may seem somewhat subjective, it is adaptable and easily performed. Since this portion of the study is examining perception of landscape similarity, it is of interest what attributes of a landscape, as defined by landscape designers, can be related to the personal constructs used by "landscape users" to perceive the landscapes.

Preference Data

The preference analysis is designed to determine which elements of the man-made environment are preferred. It presumes that preference is a complex process that can be represented as a multidimensional preference space. Since preference is often a subject of disagreement among persons with differing cognitive sets, the preference of residents for landscapes in their small town is studied. Residents of a small, somewhat isolated community are more likely than most populations to have similar cognitive sets in terms of preference for a "townscape".

STUDY AREA AND PHOTOGRAPHIC REPRESENTATION

The "townscape" of Wingham, Ontario was selected for this study of landscape preference. Wingham is a small town of 3,000 persons located in Huron County, 115 kilometres north of London, Ontario. It is an older agricultural centre surrounded by mixed farming and located on the Maitland River. The town is structurally typical of the small agricultural settlements founded in the 1800's in southern Ontario.

Forty 35 millimetre colour photographs comprising residential, commercial, recreational and mixed land uses within Wingham were taken on a sunny day in late February. The town was snow-covered at the time, characterizing winter conditions in Ontario. Photographs were taken in winter to emphasize the structural attributes of the town, rather than gardens, flowerbeds, etc., in eliciting preference. The photographs show the maximum context available, in ordinary views from the sidewalks and streets of Wingham. The ratio of sky to land in each photograph is constant at 1:5

TESTING PROCEDURE

Forty residents of Wingham were enlisted from church groups, the high school and users of the public library. Males and females are equally represented among the participants. Forty-eight percent have attended university and the work experience of the participants includes general labour, office or sales work and teaching. The majority were raised and have recently resided outside of cities, and report that they are familiar with the Wingham area. Their similar backgrounds and present choice of residence in Wingham may mean that the respondents have similar personal constructs of preference for the Wingham townscape. The participants were given a randomly mixed stack of the 40 ten by fifteen centimetre colour prints of Wingham and told that this is a study of preference for the landscapes of Wingham. They were told to rate each photograph on a preference scale of one to five, one being not preferred and five being highly preferred. The participants were then asked to list one word describing each photograph. The ranking procedure used here is faster to perform than the sorting procedure used in the perception analysis. Ranking of photographs based on preference is relatively easy, since determination of preference is an evaluation. This contrasts with determination of similarity which is based on a discrimination process and, therefore, more easily represented by a sorting procedure.

LANDSCAPES ATTRIBUTES

The photographs were inspected for several visual attributes, age of residences, relative income of neighbourhoods, presence of coniferous trees, presence of deciduous trees, openness of view, ground cover, age of businesses, skycolour, blight, presence of public buildings, presence of flags, brightness of the scene and the presence of water bodies. These attributes were selected on the basis of use in the landscape architecture profession,[26] aesthetic factors proposed by previous researchers[27] and prominence in the photographs. The gradation of an attribute over the landscapes was noted and each photograph assigned a "plus", "minus" or "zero" depending on the range of the attribute displayed by the photograph. A plus or minus was assigned to extreme gradations of attributes while a zero was assigned when the attribute did not apply or was indeterminate.

ANALYSIS

The mechanics of analysis differ somewhat between the preference and perceived similarity data. This is because of inherent differences between the procedures used to generate the repertory grids for each individual and the natures of the multidimensional spaces.

Repertory Grid Analysis

To create a repertory grid of similarity data, a binary matrix was produced for each individual, with the rows and columns representing the photographs. Each cell of the matrix has a value of one if the photographs were sorted into the same pile and zero if they were not. These similarity matrices were added together to form a single aggregate similarity matrix whose cells' values represent the degree to which a pair of landscapes are perceived to be similar by the study participants. This matrix is the repertory supergrid of perceived similarity.

To create a repertory supergrid of preference data a matrix, with the rows and columns representing the photographs, was produced for each individual. The difference between the preference ratings for each pair of photographs was calculated. These differences between preference comprise the elements of the repertory grid. The preference matrices were added together, forming a single aggregate matrix describing the differences in preference for landscapes. This matrix is the repertory supergrid of preference.

The perceived similarity and preference supergrids were analysed by non-metric alternating least squares multidimensional scaling,[28] using the Statistical Analysis System computer analysis package.

A three-dimensional MDS solution is most suitable for the similarity space. For a three-dimensional space, Kruskal's Stress = 0.141 with a much higher stress for two dimensions and little improvement for four dimensions. The $r^2 = 0.849$ for the three-dimensional solution, indicating 84.9% of the disparities between the landscapes are explained by the location of the landscapes in the three-dimensional space. The three-dimensional configuration shows a fairly wide spread of landscape "points" with a tendency for landscape points to cluster in the periphery of the space.

For the preference space a two-dimensional MDS solution is most suitable. For two dimensions, Kruskal's Stress = 0.199; there is little

Table 1,7 CENTROID VALUES OF CLUSTERS ON THE THREE DIMENSIONS

Cluster	Dimension		
	1	2	3
1.	−0.14	0.22	−1.22
2.	0.04	−1.6	−0.36
3.	−1.26	−1.14	0.57
4.	−1.36	0.67	0.49
5.	1.42	0.39	−0.05
6.	1.66	−0.16	1.14

improvement for higher dimensional solutions. The $r^2 = 0.955$ for the two-dimensional solution indicating that 95.5% of the variance between the preferences for landscapes is explained by the location of the landscapes in the two-dimensional space. The landscape points in this space have a heavy concentration in the centre of the configuration with only a few points scattered in the periphery.

Interpretation of the Landscape Similarity Space

A major intention of this study is to identify the dimensions of perceived similarity in terms of permanent attributes of the landscape. Since this configuration of landscapes in the similarity space shows a tendency for clustering it was determined that the dimensions should be interpreted in terms of landscape attributes common to each cluster.

Johnson's hierarchical clustering algorithm was applied to the perceived similarity supergrid; the derived clusters of photographs being plotted on the three dimensional similarity space. Six clearly defined clusters emerge, the centroids of which determine which particular clusters have diagnostic values for use in interpreting the identity of the dimensions. Values of the centroids range from ±0.04 to ±1.66 in the various dimensions (see Table 1,7). Values greater than 0.50 are diagnostic, those less than 0.50 are non-determinate (0) or given a positive or negative magnitude on each dimension. Magnitudes were assigned on the following basis: magnitude 1 equals ±0.5 to ±1.0; magnitude 2 equals ±1.01 to ±1.5; magnitude 3 is greater than 1.5 (see Table 2,7).

Table 2,7 INTEGRAL MAGNITUDES OF THE CLUSTER CENTROIDS FOR THE THREE DIMENSIONS
(0 = Nondeterminate Centroid)

Cluster	Dimension		
	1	2	3
1.	0	0	−2
2.	0	−3	0
3.	−2	−2	+1
4.	−2	+1	0
5.	+2	0	0
6.	+3	0	+2

Photographs with extreme values on a particular dimension are assembled into "extremes value groups". Such a group only characterizes a positive or negative range of a dimension. Extreme value groups were used only when there was no diagnostic cluster for that range of a dimension.

If the photographs in a cluster or extreme value group were consistent in their ranges of landscape attributes, then the cluster or group received that rating (e.g. positive, negative). If the landscape attributes were inconsistent the cluster or group received a non-determinate rating (0) for that landscape attribute (see Table 3,7).

Interpretation of the Landscape Preference Space

The preference space requires greater subtlety in interpretation than the perceived similarity space. The variations of preference among landscapes may be related to intangible concepts rather than easily identifiable attributes of the landscapes. The preference space may also be warped by inclusion of some aspects of similarity among landscapes. The lack of clusters in the preference space requires use of landscapes with extreme values in particular dimensions rather than clusters of landscapes to characterize attributes of the dimensions.

Table 3,7 CONSISTENT RIVERSCAPE ATTRIBUTES OF THE CLUSTERS
(0 = Nondeterminate Rating)

Cluster #	Colour	Angle of View	Vegetation	Clarity	Land Use	Valley Slope	Soil Exposure	Blight	River Exposure	Cultural Features
	Brown + Green −	High + Low −	Lush + Barren −	Clear + Hazy −	Rural + Urban −	Steep + Level −	Great + Minor −	Predom. + None −	Great + Minor −	Positive + Negative −
1.	+	0	−	+	0	0	+	0	0	0
2.	0	0	−	+	0	0	+	+	0	−
3.	0	−	−	−	−	0	+	+	0	−
4.	0	0	0	0	−	0	−	−	0	+
5.	0	0	+	+	+	+	−	−	0	0
6.	−	−	+	−	+	0	−	−	+	0

AVERAGE PREFERENCE MAPPING

To determine the variation of preference within the preference space the ratings of preference by the 40 study participants were averaged for each photograph. These mean preferences were mapped onto the two-dimensional preference space to confirm that the space adequately represents preference and to observe the variation of the mean preference within the space.

EXTREME VALUE GROUPING

Since clusters are non-existent in the periphery of the two-dimensional preference space, landscapes with extreme values in only one dimension were selected. A group of extreme value landscapes for each dimension was selected in this manner. These landscapes also possess the highest levels of mean preference. A group of landscapes clustered in the centre of the preference space with the lowest levels of mean preference was also assembled.

DESCRIPTION ASSOCIATION

To determine the possibly intangible concepts associated with landscape preference the one word descriptions given by each participant in the study were assembled for each landscape in the extreme value groupings. The words most often used to describe the landscape in each group are listed in Table 4,7.

ATTRIBUTE ASSOCIATION

To determine the landscape attributes associated with the dimensions of preference and with lack of preference, the attribute ratings given individual landscapes were compared within an extreme value group. If the range of an attribute was fairly consistent within the group, then that range of the attribute was assigned to the group. If the landscape attributes did not apply or were inconsistent, a non-determinate rating (0) was given (see Table 5,7).

RESULTS

Similarity Space

The various dimensions of the perceived similarity space were identified using the common attributes of the clusters located in the space.

Table 4,7 THE MOST COMMON ONE WORD DESCRIPTIONS FOR THE LANDSCAPES IN THE EXTREME VALUE GROUPS

Group	Descriptors
Not preferred	messy, undesirable, run-down, dull
Dimension 1 +	modern, functional, trees
Dimension 1 −	historical architecture, pleasant
Dimension 2 +	parklike, open space, play activities
Dimension 2 −	trees, natural, historical, Victorian, elegant, variety

SIMILARITY DIMENSION ONE

In dimension one, clusters three and four have strong negative ratings while clusters five and six have strong positive ratings. Clusters in the negative range of dimension one have consistent man-made and urban attributes while clusters in the positive range of dimension one have natural riverscapes with very few human intrusions. It is suggested that this similarity construct of the South Saskatchewan River Valley at Saskatoon be labeled *Degree of Development* and its gradation labeled *Natural vs. Man-Made* (see Figure 1,7).

SIMILARITY DIMENSION TWO

In dimension two, clusters two and three have strong negative ratings while the dimension two positive extreme value group displays strong positive ratings. All landscapes with high positive or negative values of dimension two possess man-made attributes. Cluster two and three consistently display trash, construction sites, vandalism and utilitarian styles of architecture. The positive extreme value group displays the city centre riverscape with arching bridges, urban parks and old world styles of architecture. There is no trash or vandalism evident in the dimension two positive extreme value group. It is suggested that this similarity construct of the South Saskatchewan River Valley at Saskatoon be labeled *Structural Development Variation* and its gradation labeled *Blight vs. Enhancement* (see Figure 1,7).

SIMILARITY DIMENSION THREE

In dimension three, cluster one has strong negative ratings, while cluster six has strong positive ratings. Cluster one displays consistently brown

Table 5,7 CONSISTENT ATTRIBUTES OF EXTREME VALUE GROUPS OF LANDSCAPES IN THE WINGHAM PREFERENCE SPACE (0 = Nondeterminate Rating)

Group	Residential District	Income of Residents	Pine Trees	Deciduous Trees	View	Ground Cover	Business District	Sky Colour	Blight	Public Structures	Canadian Flags	Brightness of Scene	Water Bodies
	+ Old – New	+ Upper – Lower	+ Prevalent – None	+ Open – None	+ Shrubs – Closed	+ Old – Snowfields	+ Blue – New	+ Prevalent – White	+ Prevalent – None	+ Prevalent – None	+ Light – None	+ Prevalent – Dark	– None
Not Preferred	0	0	+	+	0	0	0	+	0	–	0	–	
Dimension 1+	–	+	0	+	–	0	–	+	–	0	0	–	–
Dimension 1–	+	+	0	0	+	0	+	+	–	+	0	0	–
Dimension 2+	–	0	0	+	+	–	0	+	–	–	–	+	0
Dimension 2–	+	+	+	+	+	+	0	0	–	–	–	–	0

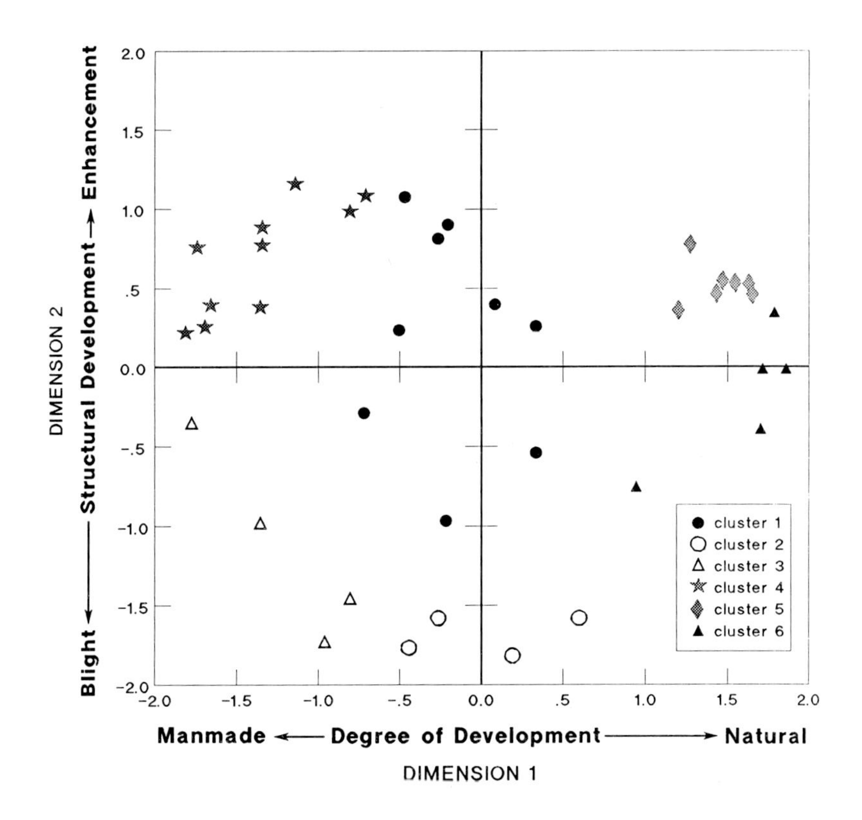

FIGURE 1,7 Saskatoon Riverscape Similarity Space

coloured land, barren vegetation and often disturbed soils. Cluster six displays a lush, green landscape of thick vegetation and no bare soil. Landscapes with extreme values of dimension three have little in the way of man-made attributes or interference. It is suggested that this similarity construct be labeled *Natural Variation* and its gradation be labeled *Barren and Brown vs. Lush and Green* (see Figure 2,7).

Preference Space

The two dimensions of the preference space were identified in terms of the words used to describe, and common attributes of, the landscapes

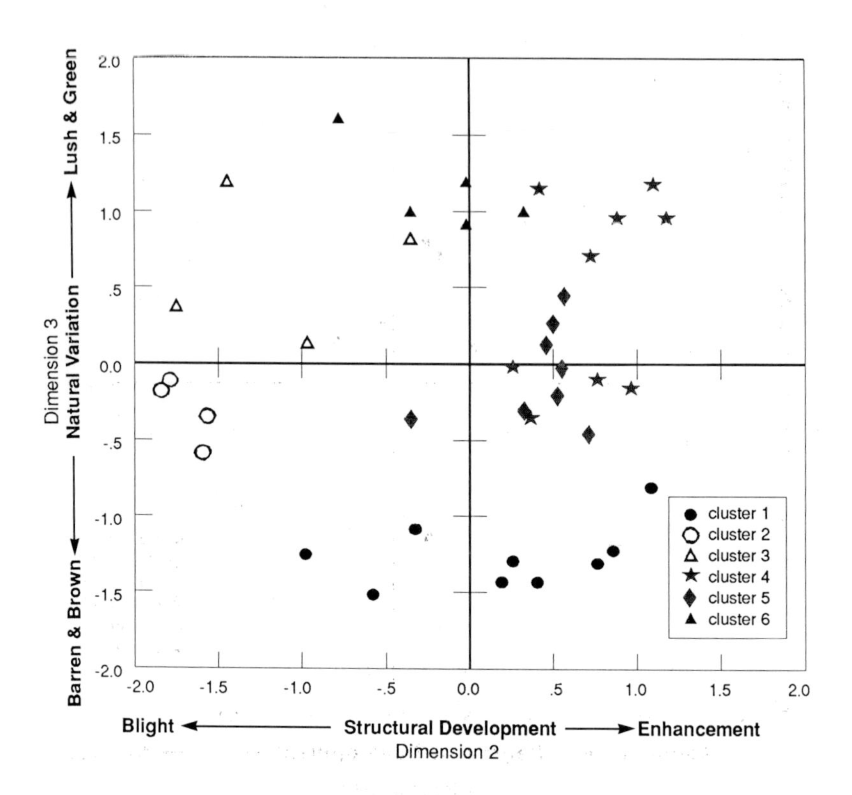

FIGURE 2,7 Saskatoon Riverscape Similarity Space

possessing extreme values in one of the dimensions. The word descriptions and landscape attributes tend to concur, though the information carried about the landscapes is often complimentary.

PREFERENCE DIMENSION ONE

Landscapes with strong values in dimension one display upper income residences and well-kept business districts; natural vegetation is of little importance to this dimension. The positive range of dimension one displays new residential and business districts with no blight evident. These landscapes are described as "modern" and "functional". The negative range of dimension one displays older Victorian residences, churches and war monuments; there is no blight evident. These landscapes are described as "historic architecture" and "pleasant". It is, therefore, suggested that this preference construct be labeled *Businesses,*

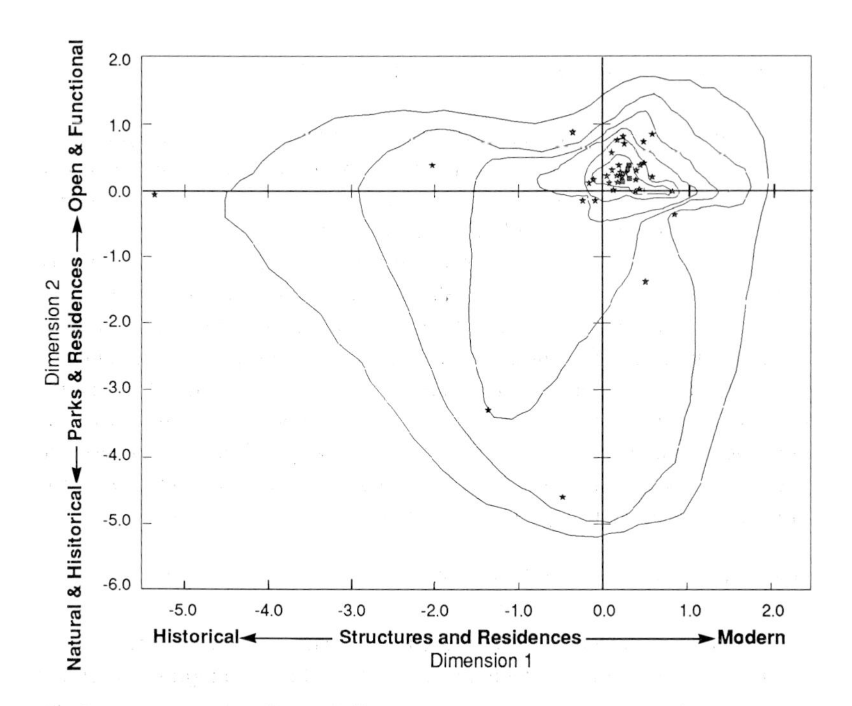

FIGURE 3,7 Wingham Landscape Preference Space

Public Structures and Residences with the gradation labeled *Modern vs. Historical* (see Figure 3,7).

PREFERENCE DIMENSION TWO

Landscapes with strong values in dimension two exhibit medium to upper income residences, deciduous trees, open views, no business district, occasional views of the river and frequent views of parks. There is no blight evident. The positive range of this dimension shows newer residences, no pine trees, deciduous trees, open snowfields and developed parks. These landscapes are described as "parklike", "open space" and "play activities" areas. The negative range shows older residences, pine trees, deciduous trees, more vegetated ground cover and relatively undeveloped parks, described as "natural", "treed", "historical", "elegant" and "Victorian". It is suggested that this preference construct be labeled *Parks and Residences* with the gradation labeled *Open and Functional vs. Natural and Historical* (see Figure 3,7).

MEAN PREFERENCE WITHIN THE PREFERENCE SPACE

The variation of mean preference within the two dimensional preference space is demonstrated by a mapping of isolines of preference (isoprefs) onto the preference space (Figure 3,7). Note that the space is not Euclidean in terms of mean absolute preference. This is to be expected, as the preference space provides a deeper insight into the complex nature of preference than a simple bipolar scale. The isoprefs increase in an approximately concentric pattern from the centre of the space. In other words, as the magnitude of a dimension increases in either the positive or negative range, the mean preference increases as some function of the dimensional magnitude. The functions of the increase of mean preference with the dimensional magnitudes are plotted for each range of both dimensions in Figure 4,7. Note that the rate of increase of mean preference with the dimensions varies among the dimensions.

NON-PREFERRED LANDSCAPES

Landscapes with mean preferences below two (on a scale of one equals not preferred, five equals highly preferred) are plotted in Figure 3,7 on the two dimensional preference space. Landscape attributes and one word descriptions have been assembled for this group of landscapes in the same manner as the other preference groups. These landscapes are described as "messy", "undesirable", "rundown" and "dull". Their consistent landscape attribute is the presence of blight in some form. The blight consists of open junk yards, abandoned buildings and utility stations within the town. None of the blighted landscapes are in upper income residential areas, most being in commercial areas or back alleys. These landscapes are located in the approximate centre of the preference space, having low values in all the dimensions.

DISCUSSION

In the case of the perception of similarity of the South Saskatchewan River Valley at Saskatoon, the participants were evaluating the differences between riverscapes in and near Saskatoon. The cognitive set within which these landscapes are evaluated is determined by the variation of landscape attributes in the area studied. The important landscape attributes have been defined in terms of the common constructs of the similarity space. While the importance of the attributes and constructs is specific to the "landscape region" of the Saskatoon riverscape, they are potentially very useful tools in regional planning of landscapes.

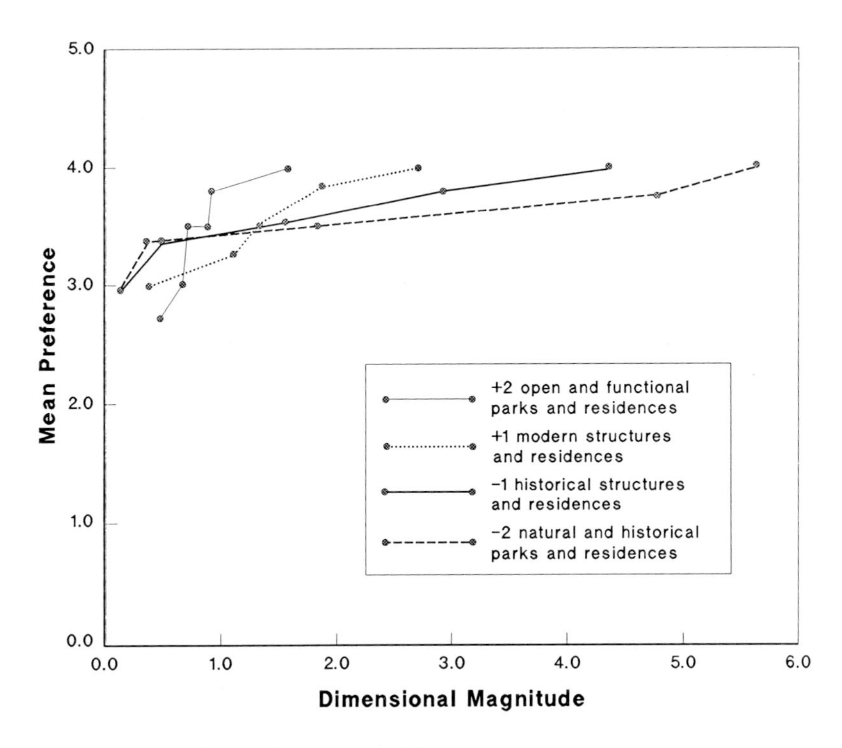

FIGURE 4,7 Preference for Dimensional Ranges

The *Degree of Development* similarity construct can be used to determine the apparent "naturalness" of an environment and to indicate what landscape attributes result in that perception. Landscapes perceived as natural may differ greatly from ecologically undisturbed environments.[29] This distinction is important in land management, particularly where there are pressures for natural landscapes in areas where establishment of such reserves is impossible due to land use conflicts. The impact of new developments in "natural areas" may be assessed using this construct as well.

The *Structural Development Variation* similarity construct evaluates riverbank development in the context of the structures along the South Saskatchewan River. This similarity construct seems to be strongly influenced by preference for specific types of structural development. However, the sample of built environments along the Saskatoon riverbanks is somewhat limited in terms of the types of structures found in the city. The direct views of the Saskatoon city centre include arching bridges and the Hotel Bessborough, acknowledged city landmarks. These views have the highest positive values on Dimension 2. Views showing the steel traffic bridge and the skyline of modern apartment buildings rate somewhat lower on the *Enhancement* range of the *Structural Development Variation* dimension. Views showing construction areas and trash with the steel railway bridges rate high on the *Blight* range of the *Structural Development Variation* dimension. This dimension seems to be based on preference; however, preference was not specifically elicited in the similarity study. It is therefore suggested that only the elements of preference important to judgements of the similarities among the landscapes compose this dimension. The dimension displays the differences perceived to be important among the structures on the Saskatoon riverbanks; preference for the structures is an element of these differences.

The *Natural Variation* construct evaluates the land use, park development, moisture regime, soil erosion and dominant vegetation attributes of riparian lands. It does not evaluate man-made structures but does respond to unnatural disturbances of vegetation and soil. "Natural" components of the landscape are qualified on a perceived vegetation-colour scale. If it can be assumed that lush and green landscapes are preferred to barren and brown ones, then an element of preference for natural vegetation may influence this dimension. This construct has potential for evaluating the perceived aesthetic impact of riverbank erosion, park designs, trail locations, landscaping of buildings, agricultural land use and soil moisture supply. Specific patterns and types of vegetation can be evaluated and suggestions for effective manipulation of vegetated landscapes can be made on this basis.

The two preference constructs found for the townscape of Wingham, Ontario define a space with the lowest preference in the centre and increasing preference away from the centre. This contrasts with the *Structural Development Variation* and *Natural Variation* constructs of Saskatoon's similarity space where low preference is confined to one range and high preference to the opposite range of the dimensions. The vast majority of landscapes from Wingham are concentrated in the low preference centre of the space, with a few outstanding landscapes around the periphery of the space. The preference dimensions define the attributes and values of landscapes that cause them to be highly preferred.

The *Businesses, Public Structures and Residences* construct of the preference space identifies landscapes containing those features that are preferred because they are modern and functional and those preferred because they are historical and pleasant. The *Parks and Residences* preference construct identifies landscapes preferred because they are considered open space and good recreational areas and those preferred because they are considered natural, elegant and historical. Thus the features of a landscape that cause it to be preferred can be identified and compared to features in other landscapes. The landscape with the highest preference in

Wingham is on the *Historical* range of the *Businesses, Public Structures and Residences* construct and displays a stone war monument, an old church and a residence from the 1800's. The scene represents the best features of 19th century architecture in Wingham and is therefore outstanding in the context of the Wingham landscape.

Figure 4,7 indicates that the greatest response of preference is to the *Open and Functional* range of the *Parks and Residences* construct. A small increase in this attribute of a landscape will cause the greatest increase in preference among the various ranges of the dimensions. Preference has its second greatest response to the *Modern* range of the *Businesses, Public Structures and Residences* construct. The *Historical* range of this construct elicits the third greatest response. The *Natural and Historical* range of the *Parks and Residences* construct generates the lowest response of preference increase to an increase in dimensional attributes.

This indicates that preference is most sensitive to development of open, functional and modern townscapes. However, the least preferred landscapes also have these attributes to some degree. Inclusion of trash or rundown buildings in these types of landscapes causes a dramatic drop in preference. Thus, in Wingham, an "average" landscape can have its level of preference increased most easily by adding attributes of open space, functional activities and modern housing. However, this is the most unstable landscape in terms of preference and the easiest landscape to blight. Within Wingham these landscapes have not elicited the highest preference, being only moderately to well preferred.

Preference is least sensitive to changes in the historical and natural landscapes. These townscapes are not being produced anymore in Wingham, as are the open, modern and functional townscapes. However, the historical and natural areas are difficult to degrade. Major variance in their attributes cause little change in preference. These landscapes are capable of sustaining the highest preference in Wingham. Historical and natural landscapes in this small town should therefore be of great interest to local planners, as they are capable of developing the highest degree of preference but when blighted, require large changes in their attributes to restore. Old and natural landscapes already possessing high degrees of preference are therefore rare and to be valued.

CONCLUSION

The stress and r^2 values of the multidimensional spaces indicate that the application of personal construct theory using the repertory grid and MDS can be successful in both similarity and preference studies of landscape perception. The dimensions or "constructs" of similarity and preference found in this study are found to function within the experimentally specified cognitive set of the study participants. This allows evaluation of landscapes in a regional context. Thus, each evaluation of landscapes can be tailored by the experimenter to a region or aspects of a region which are of interest.

The similarity space can be useful in designing landscape changes which do not significantly alter perceptions of the landscapes reserved for preservation. Preference is found to be a component of some dimensions of the similarity space, particularly those evaluating landscapes which have been altered by man. The similarity space defines the attributes of an area that are important to some local residents. Wider testing of the perceptions of various socio-economic and cultural groups may show that not all attributes of a landscape are perceived equally by various groups.

The preference space configuration confirms that residents from both sexes, various age groups and income levels can have similar preferences for their hometown landscapes. The preference space easily isolates outstanding landscapes and determines the qualities of these landscapes that make them outstanding. The sensitivity of preference to changes in various landscape qualities is determined as well.

Similarity and preference spaces open up many new possibilities in assessing the aesthetics of landscapes. Quantification of generally perceived similarity and preference allows aesthetic impact assessments to be performed as well as monitoring the aesthetic qualities of an area or site. Such relatively exact measurements of aesthetics may allow "tailoring" of landscapes to specific user groups. The principles which have guided structural architecture, landscape architecture and regional planning may be rigourously tested. The scientific and theoretical soundness as well as the empirical viability of the methodologies espoused by this paper will allow the aesthetics of landscape to have a greater input in land planning and management decisions.

REFERENCES

1. KELLY, G.A., *The Psychology of Personal Constructs*. New York: W.W. Norton and Company, 1955.

2. BANNISTER, D. and FRANSELLA, F., *Inquiring Man: The Theory of Personal Constructs*. Harmondsworth, U.K.: Penguin, 1971; O'HARE, D.P. and GORDON, I.E., "An application of repertory grid technique to aesthetic measurement", *Perceptual and Motor Skills*, 42, 1976, pp. 1183-1192.

3. HARRISON, J. and SARRE, P., "Personal construct theory, the repertory grid and environmental cognition", in MOORE and GOLLEDGE (eds.), *Environmental Knowing: Theory, Research and Methods*. Stroudsburg, Pa.: Dowden, Hutchinson and Ross, 1976, pp. 375-384; O'HARE, D.P., "Individual differences in perceived similarity and preferences for visual art: a multidimensional scaling analysis", *Perception and Psychophysics*, 20, 1976, pp. 445-452; ULLRICH, J.R. and ULLRICH, M.F., " A multidimensional scaling analysis of perceived similarities of rivers in western Montana", *Perceptual and Motor Skills*, 45, 1976, pp. 575-584.

4. HARRISON J. and SARRE, P., *op. cit.*

5. O'HARE, D.P., *op. cit.*; O'HARE, D.P. and GORDON, I.E., *op. cit.*

6. GARLING, T., "The structural analysis of environmental perception and cognition: a multidimensional scaling approach", *Environment and Behavior*, 8, 1976, pp. 385-415.

7. HARRISON, J. and SARRE, P., *op. cit.*; WARD, L. and RUSSEL, J., "Cognitive set and perception of place", *Environment and Behavior*, 13, 1981, pp. 610-632.

8. ULLRICH, J.R. and ULLRICH, M.F., *op. cit.*

9. JUTLA, R.S., "An Evaluation of Visual Perception and Preferences of a Townscape", Unpublished Master's Thesis, University of Guelph, 1983; PEARCE, S.R. and WATERS, N.M., "Quantitative methods for investigating the variables that underlie preferences for landscape scenes", *Canadian Geographer*, 27, 1983, pp. 328-344; POMEROY, J.W., GREEN, M.B. and FITZ-

GIBBON, J.E., "Perception of non-spectacular landscapes: a comparison of methodologies", *Research Report, University School of Rural Planning and Development*, Guelph: University of Guelph, 1982; POMEROY, J.W., GREEN, M.B., and FITZ-GIBBON, J.E., "Evaluation of urban riverscape aesthetics in the Canadian prairies", *Journal of Environmental Management*, 17, 1983, pp. 263-276.

10. O'HARE, D.P., *op. cit.*

11. *Ibid.*

12. WARD, L. and RUSSEL, J., *op. cit.*

13. O 'HARE, D.P., *op. cit.*

14. YOUNG, F., TAKANE, Y. and LEWYJKYJ, R., *ALSCAL-4 User's Guide*, Carrboro, N.C.: Data Analysis and Theory Associates, 1979.

15. HARRISON, J. and SARRE, P., *op. cit.*; PEARCE, S.R. and WATERS, N.M., *op. cit.*; POMEROY, J.W. *et al.*, *op. cit.*, 1983; ULLRICH, J.R. and ULLRICH, M.F., *op. cit.*; WARD, L. and RUSSEL, J., *op. cit.*

16. HARRISON, J. and SARRE, P., *op. cit.*

17. WARD, L. and RUSSEL, J., *op. cit.*

18. DEARDEN, P., "A statistical method for the assessment of visual landscape quality for land use planning purposes", *Journal of Environmental Management*, 4, 1976, pp. 15-26.

19. BRUSH, R. and SHAFER, E., "Application of a landscape preference model to land management", in ZUBE, E. *et al.* (eds.), *Landscape Assessment: Values, Perceptions and Resources*. Stroudsburg, Pa.: Dowden, Hutchinson and Ross, 1975, pp. 168-182; BUYHOFF, G. and WELLMAN, D., "The specifications of a non-linear psychophysical function for visual landscape dimensions", *Journal of Leisure Research*, 12, 1980, pp. 257-272; CALVIN, J., DEARINGER, J. and CURTIN, M., "An attempt at assessing preferences for natural landscapes", *Environment and Behavior*, 4, 1972, pp. 447-470; DUNN, A., "Landscape with photographs: testing the preference approach to landscape evaluations," *Journal of Environmental Management*, 4, 1976, pp. 15-26; FINES, K., "Landscape evaluation: a research project in

east Sussex", *Regional Studies*, 2, 1968, pp. 41-55; PEARCE, S.R. and WATERS, N.M., *op. cit.*; POMEROY, J.W. *et al.*, *op. cit.*, 1982, 1983; ZUBE, E., PITT, D. and ANDERSON, T., "Perception and prediction of scenic resource values of the Northeast", in ZUBE, E., *et al.* (eds.), *Landscape Assessment: Values, Perceptions and Resources*. Stroudsburg, Pa.: Dowden, Hutchinson and Ross, 1975, pp. 151-169.

20. SHUTTLEWORTH, S., "The use of photographs as an environment presentation medium in landscape studies", *Journal of Environmental Management*, 11, 1980, pp. 61-76.

21. DUNN, A., *op. cit.*

22. POMEROY, J.W., "The Influence of River Regime on Recreation in the South Saskatchewan River", Unpublished College of Arts and Science Scholar Thesis, University of Saskatchewan, 1983.

23. LEOPOLD, L., "Quantitative comparisons of some aesthetic factors among rivers", *U.S. Geological Survey Circular*, 620, Washington, D.C.: U.S. Department of the Interior, 1969.

24. LINTON, D., "The assessment of scenery as a natural resource", *Scottish Geographical Magazine*, 84, 1968, pp. 219-238.

25. LITTON, B., TETLOW, R., SORENSON, J. and BEATTY, R., *Water and Landscape: An Aesthetic Overview of the Role of Water in the Landscape*. Port Washington, N.Y.: Water Information Centre, 1974.

26. United States Department of Agriculture, "Procedure to establish priorities in landscape architecture", *Technical Release no. 65*, Washington, D.C.: United States Department of Agriculture, Engineering Division, Soil Conservation Service, 1978.

27. GARLING, T., *op. cit.*

28. YOUNG, F. *et al.*, *op. cit.*

29. FITZGIBBON, J.E., O'HARE, T.S., POMEROY, J.W. and RICHARDS, G., *A Baseline Study of Meewasin Valley River Resources*. Saskatoon: Meewasin Valley Authority, 1982.

PLATE 16 Heather Mountain, British Columbia. Land use activities influence not only scenic quality but also landscape processes: logging often results in landslips, particularly below road-cuttings. ►

8 ENVIRONMENTAL AND POLICY REQUIREMENTS: SOME CANADIAN EXAMPLES AND THE NEED FOR ENVIRONMENTAL PROCESS ASSESSMENT

Michael R. Moss and William G. Nickling

INTRODUCTION

Over the past two decades, there has been an increased awareness of the importance of landscape analysis or scenic assessment in our overall assessment of the environment. Advancements in this area of enquiry, particularly in methodolgy, are similar to developments in other research areas which collectively contribute to the field of environmental management. The first stage can usually be identified by an increasing awareness of a particular environmental problem from the perspective of existing management and policy frameworks. The second stage may often be characterized by an increasing domination of concerns about management and policy-related issues. Once curiosity has been aroused by the scientific community or the general public, any further developments from this perspective tend to decrease proportionately as concern with policy increases. These, and similar trends, can be detected in such sub-fields of environmenal management as impact assessment, evaluation of the biophysical or ecological bases for land planning, and in assessments of natural and environmentally sensitive areas. In each case, advances in strategic planning and policy-related analysis have far outstripped equivalent advances in our understanding of the environmental bases of the problems themselves. Only from an increased understanding of the appropriate environmental systems, however, come sound management strategies.

The field of landscape or scenic assessment is no exception to the generalizations outlined above. One independent development within this field is that in the second stage, concern has tended to focus upon an understanding of "how" and "why" people perceive landscapes as they do. Developments in this area have tended to overshadow concerns with policy and related issues. However, the common thread in each case is that the attention of the resource planner is diverted away from the critical issues, that is from an understanding of the environmental bases and a consequent failure to develop more effective implementation strategies within a planning context. In this paper, a brief examination will be made initially of the existing policy requirements for incorporating information on scenic and related landscape properties into land planning decisions within Canada. Subsequently, several techniques of landscape assessment (i.e. the scenic value of land) will be discussed, and the degree to which they use appropriate environmental information analysed. A final section deals with the need to improve the entire system of land data acquisition and points out areas where improved approaches will enhance our capacity to plan environments more effectively, and to link related techniques of environmental analysis — of which landscape assessment is only one — to developments of process-related rather than inventory-related data acquisition.

In discussing such problems, possible developments in the field of landscape aesthetics should be isolated from those in other fields of environmental management. Effective connections need to be made between these interrelated aspects of environmental assessment. Mitchell[1] points to three main areas where landscape evaluation provides links with other types of environmental assessment needed for effective environmental planning:

1) improvements in resource inventories,
2) in relationship to carrying capacity decisions, and
3) associated with environmental impact assessment.

These inter-connections have been developed in association with required improvements in the biophysical basis of environmental assessment, particularly where this relates to land potential and its link with carrying capacity determination and environmental impact assessment.[2]

A major theme which emerges from this work has been the need to recognize and establish procedures for the use of techniques which employ information on land process attributes. Resource inventory as the sole base data input can have severe limitations for effective use in environmental planning.[3] Practitioners on both sides of the issue may be at fault. The social scientist in the field of resource management is often ignorant of the limitations of the basic environmental data with which he is presented. In contrast, the environmental scientist frequently ignores the need to produce the appropriate data in a format suitable for use by the resource manager.

POLICY REQUIREMENTS FOR LANDSCAPE EVALUATION FOR ENVIRONMENTAL PLANNING IN CANADA

A fundamental question relates to what particular level(s) of planning authority actually require, or permit input on landscape analysis to questions of environmental and regional planning. In Canada, few requirements or opportunities exist for input of landscape aesthetic information. During the past fifteen years, however policies and environmental impact assessment legislation have been introduced which provide opportunities for input. For example, the procedures of the Federal Environment Assessment and Review Process, introduced in 1973, include a series of matrices for screening projects.[4] In the 'level 1' matrix, recognition is given to "aesthetic effects", although not specifically to landscape or scenery as such. No methodology is suggested by which to analyse the issue but the opportunity exists for the incorporation of appropriate information.

At the regional and municipal level, opportunities also exist for the incorporation of landscape aesthetic information, though specific criteria and methods of appropriate data collection are seldom outlined. One exception to this is British Columbia where the Ministry of Forests has produced a methodological handbook based largely on the USDA system for use in designing forest harvesting procedures.[5] The Ministry of Environment has also developed procedures,[6] but as with the forest handbook these are seldom implemented. Two cases can be cited where good use of landscape information was made in Canada. These are the Banff Highway Project[7] and the plans of Ontario Hydro for major transmission corridors.[8] In each case, however, the main concern was with the visual impact of the highway or hydro corridor on the environment, rather than upon units of landscape designated as having a degree of scenic value and on which the proposed developments would have detrimental effects. It is also worth noting that the work undertaken by the Niagara Escarpment Commission in Ontario (throughout the 1970s) did not result in the designation of one area of the Escarpment as having a degree of scenic value (as opposed to areas of special ecological value). Yet, the Niagara Escarpment Commission was established as the coordinating planning authority for what is arguably Ontario's most extensive scenic landscape feature.[9]

LANDSCAPE EVALUATION USING INVENTORY DATA

The next question for consideration is whether available landscape data or the methods of collection are appropriate for input to the environmental planning process. This notion can be examined with examples at two scales: firstly, at a large scale in the Niagara Peninsula of southern Ontario, where the issues relate to environmental planning at the municipal level; and, secondly, in a small scale analysis of the Lancaster Sound area in the Arctic where ideas of establishing a northern national park have frequently been raised.

On the Niagara Peninsula, Moss and Nickling[10] tested the adaptability and efficacy of several established procedures in identifying sites of high to low scenic value. Three commonly used procedures applied were, an adaptation of the Linton method,[11] the method developed by Newkirk *et al.*[12] and the Leopold[13] system of riverscape analysis. Although there are certain procedural and methodological problems associated with each of these methods (see Hamill in this volume), each one arguably has the ability to identify areas of high or low scenic value. The Linton and Newkirk methods are particularly appropriate for regional planning purposes since mapped data can be readily obtained by both field and cartographic survey techniques.[14] An example of this type of result is shown in Figure 1,8. Similarly, the Leopold method of determining "uniqueness ratios," while being much more time-consuming and subject to some serious questions of statistical validity, nevertheless permits the ready identification of units of riverscape with varying degrees of uniqueness. In each method, the objective is to develop an inventory of resource data by spatial unit.

Although initially developed for relatively large scale studies, based upon 1 to 10 kilometre grid squares, the Linton method can be modified for use at smaller scales. In an exercise designed, in part, to identify potential areas of scenic value for a northern national park in the Lancaster Sound area of the Canadian Arctic, the method was adapted to both regional environmental criteria and to an existing data base. In this instance the bases of the spatial units are previously identified ecological or biophysical land units, rather than grid squares.

The sources are the ecoregional maps and resource inventories of the Canada Committee on Ecological Land Classification of the Lands Directorate of Environment Canada. At the ecoregional scale,[15] this resource inventory contains data on a range of environmental attributes related to landscape such as topographic variation, soil and vegetation conditions, ground water coverage and similar factors. By allocating a simple scoring system to the presence of the more positive landscape features and ranking each ecoregion on the basis of total landscape attributes, from high to low, each ecoregion can be quickly assigned a rating on a scale from high to low scenic value. Likewise, areas determined to be unique from the point of view of possessing landscape features peculiar to this region of the Canadian north can be identified. The results of these two analyses are shown in Figure 2,8. Although subjectively based, this approach permits a quick,

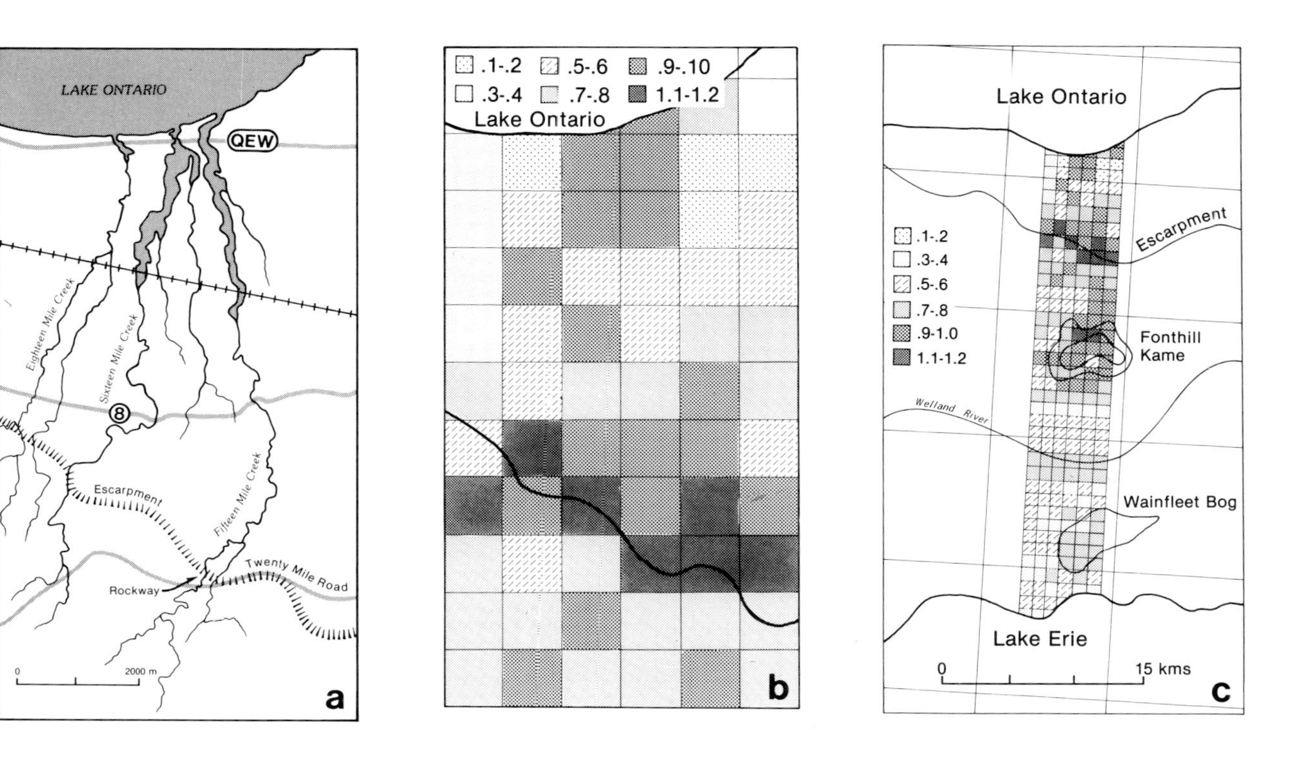

FIGURE 1,8 Examples of scenic assessment by grid square units (a) location map of part of the Niagara Peninsula (b) results of landscape analysis using the method developed by Newkirk *et al.*[12] based on 1 km. grid suares, (c) the same method based on topographic map analysis.

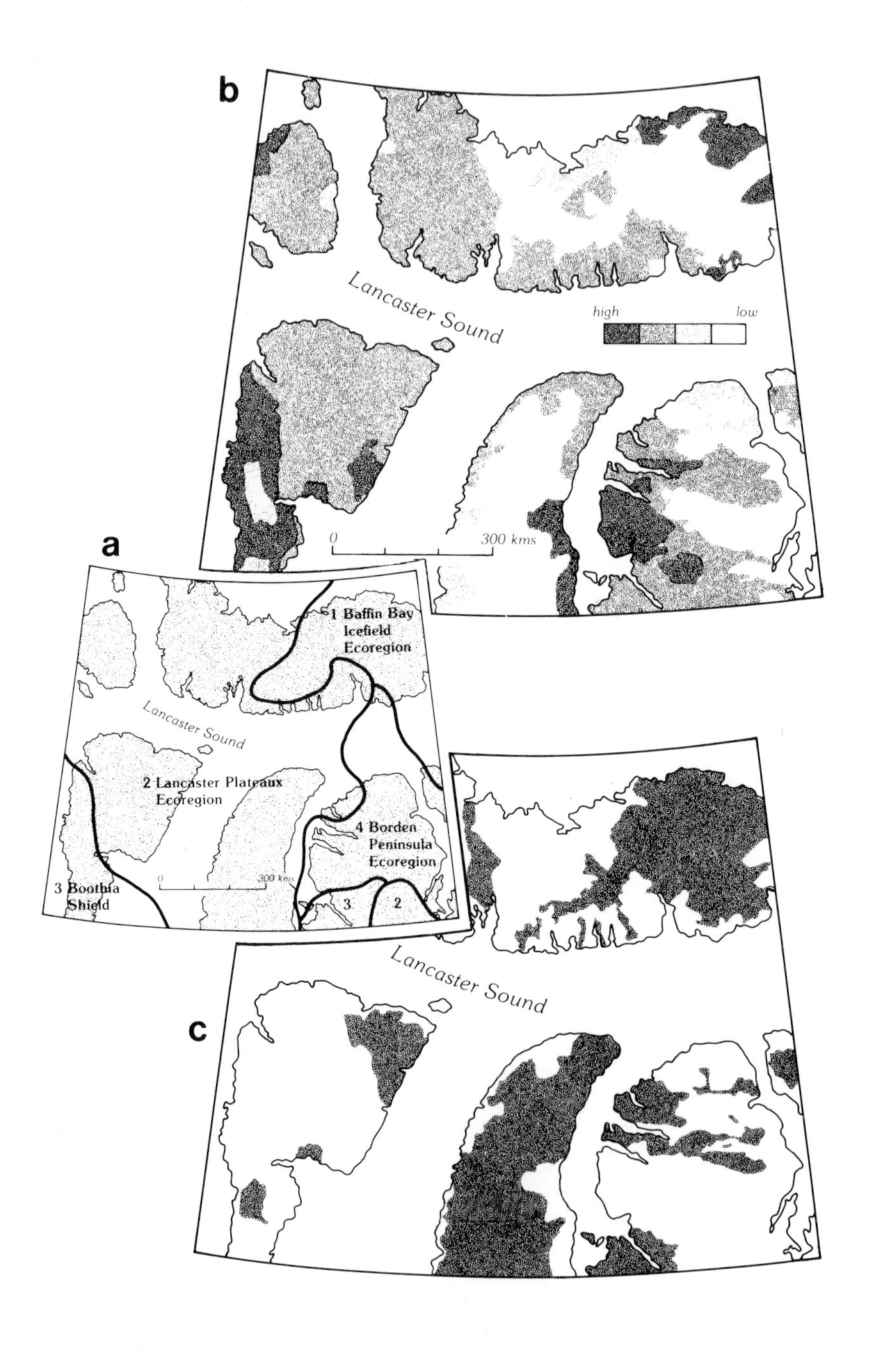

FIGURE 2,8 Examples of scenic assessment using ecological land units (a) the ecoregions, (b) relative scenic valued derived from ecodistrict information, (c) unique areas identified on the same inventory data.

environmentally based analysis to be made for determining areas of varying degrees of scenic value within an extensive region.

This type of approach, based on visible landscape attributes, has the advantage of identifying small or large scale areas possessing particular scenic resource value. But beyond saying that, further use of this type of assessment is limited. The environmental planner or resource manager is provided with neither an indication of the likely needs of particular sites within these areas nor where management strategies need to be developed to conserve areas designated to be of particular significance. This is especially so when the time-frame for environmental resource planning is on the scale of decades rather than of a few years. In order to overcome such problems, which relate also to many other aspects of environmental management, a more refined awareness of landscape dynamics needs to be developed. From this methodological refinement should develop a more logical approach to environmental data acquisition and utilization for management purposes. This problem will be discussed below by reference to a range of studies previously undertaken along, or in the vicinity of, the Niagara Escarpment of southern Ontario and based upon attempts to understand environmental dynamics.

ENVIRONMENTAL PROCESSES AND LANDSCAPE ASSESSMENT

Many studies on landscape have indicated that the higher, the more varied, and the more accentuated the topography, the greater the perceived scenic resource value of that area. This point may be supported by the dominance of cliffed seascapes, incised riverscapes, dramatic escarpments and mountainous topography in the park and reserve areas designated by many levels of government in Canada. In areas of more subdued relief, such as southern Ontario, it is the river valleys and rock outcrop areas that are more readily designated as having scenic value.

All such areas have two things in common. By using the types of schemes outlined in the previous section, areas are determined to be of high value predominantly on the basis of relief or topographic attributes. But all these areas are also "high energy" environments in the terminology of landscape dynamics. This means, therefore, that they are areas where landscape change, from an evolutionary point of view, is more likely to be fast and irregular than it is in areas where landscape processes occur at slower and more uniform rates. They are likely to be more unstable than stable environments; they are the types of landscape where development recurs on a time scale measured over a few decades rather than centuries. It is precisely upon a period of a few decades, however, that most planning strategies are focused.

A second major consideration relates to the fact that change in such relatively active environments takes place in those landscape components directly related to topography, rather than in the topography itself. Because of the nature of these processes, it is within the weathered or surficial materials and in the vegetation cover that the relevant spatial and temporal concerns with landscape dynamics are likely to have the greatest degree of expression.

In most instances, the identification of a vegetative cover is a critical criterion for scenic designation. The dynamics of this particular environmental subsystem also is subject to change and development through the time span of normal planning strategies. This argument has been developed at greater length in discussion of other resource management problems where a distinction was made between "active" and "passive" landscape components and the identification of relevant landscape processes as a focus for a more general system of classification of land for development purposes.

Reference to the significance of geomorphic processes influencing scenic quality is illustrated by reference to Figure 3,8. Figure 3,8a is a generalized map showing the range of uniqueness values derived by applying the Leopold method to parts of two relatively small river valleys flowing north from the Niagara Escarpment to Lake Ontario across the flat, tender fruit and vine-growing region of the Niagara Peninsula. The two river valleys are incised into this plain creating heavily forested valleys within an otherwise agriculturally-dominated land use pattern. As such, they form local areas of high scenic quality possesing many natural attributes. They have not yet been designated for conservation purposes. Figure 3,8b maps the nature and occurrence of mass movement and slope-forming processes encountered on these river valley sides. Such features are naturally occurring geomorphic events which for the most part will affect only restricted sections of the valley sides. Yet the cumulative impact of these features will, at one time or another, have a significant visual impact on particular locations. They also affect the vegetation, particularly the forest cover, of the valleys sides. In other words, landscape components, within the regional planning framework can be seen both as hazard control features and as scenic resources. Two issues of environmental management are thus effectively interconnected; yet no procedures exist to bring these out.

Another illustration of the need to consider vegetation dyanmics can be drawn from an area five kilometres south of the area discussed above, in what is now the Short Hills Provincial Park. This heavily wooded area on highly undulating sandy materials has many unique natural features; it contains many species of the Carolinian forest type. The original plans for the Park identified many of the forest stands as unique or characteristic forest associations which have survived in the area with little modification

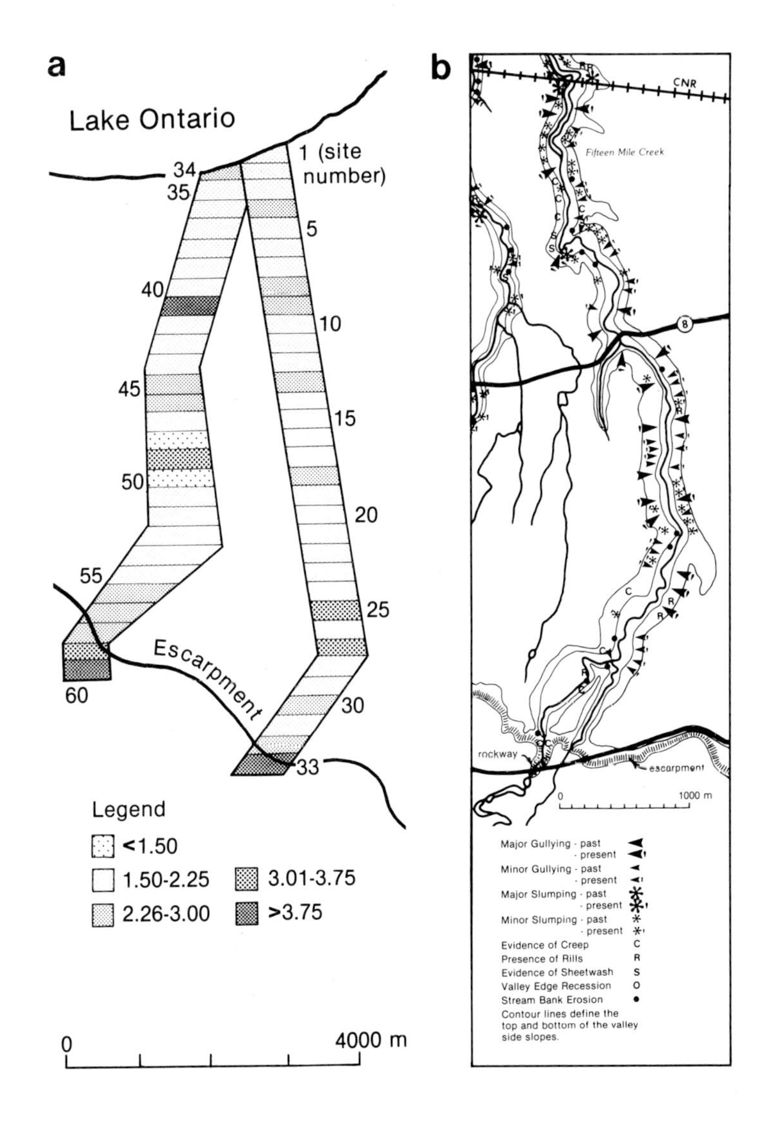

FIGURE 3,8 Fifteen and Sixteen Mile Creeks (a) uniqueness ratios derived by the Leopold method, (b) detailed mapping of hazard features along the same two river valleys. Each symbol represents a different type of active or passive geomorphic event (slump, gulley, landslip etc.).

since early settlement times. Figure 4,8 shows these forest types together with a value for each stand based on the derivation of a 'continuum index.'[17] The 'continuum index' procedure is capable of indicating the likely degree of change in forest composition over the next few decades. The closer the forest stand value is to 3000, the closer is the stand to an assumed regional climax; and, by implication, the less change is likely to occur in stand composition. Conversely, the lower the value, the more likely is significant compositional change. Figure 4,8 therefore identifies many sites likely to undergo significant natural change; in this case in the vegetation-dominated component of the landscape. Thus the components identifed as being worthy of preservation, will undoubtedly change. Since park planning in the area subsequently evolved around forested landscapes, the value of this approach lies in emphasizing the fact that landscape is dynamic and subject to change. The approach also specifically identifies those landscape features which require attention in the development of new management strategies.

The dynamics of interaction between land surface processes and forest communities can also be illustrated from a location several miles to the west, where Forty Mile Creek cuts through the Niagara Escarpment south of the town of Grimsby on the south shore of Lake Ontario. This feature is one example of the many glacial re-entrant valleys within the Escarpment. These are generally too isolated and rugged to show any major human impact and consequently many of these have been designated 'environmentally sensitive areas.' As such, they come under the jurisdiction of the regional or county governments. As scarp face features, they also come under the jurisdiction of the Niagara Escarpment Commission.

Once identified as important landscape features, however, what are the implications for future environmental planning involving these features? Figure 5,8 illustrates several process-related features of relevance: 5 (a) the nature and type of surficial mass movement; 5 (b) the dates calculated by tree ring analysis to indicate the approximate number of years since the latest occurrence of these events; and finally, 5 (c) the pattern of tree age within the gorge indicating a decrease in age up-slope as the frequency of minor disturbance events occurs.[18] As a general conclusion, it was calculated that it requires about forty years for a closed canopy to reform once opened by geomorphic mass movements, and that such events are common recurrent events worthy of recognition in any planning strategy.

A final illustration of the complexity of process interaction is shown in Figure 6,8.[19] This relates to the situation on the Niagara Escarpment in the Blue Mountain area, west of Collingwood on Georgian Bay. This is an area of both summer and winter tourist activity where the Escarpment essentially forms the *raison d'etre* for much of the economy of the region. At present, the scarp face is characterized by forest cover, open ski

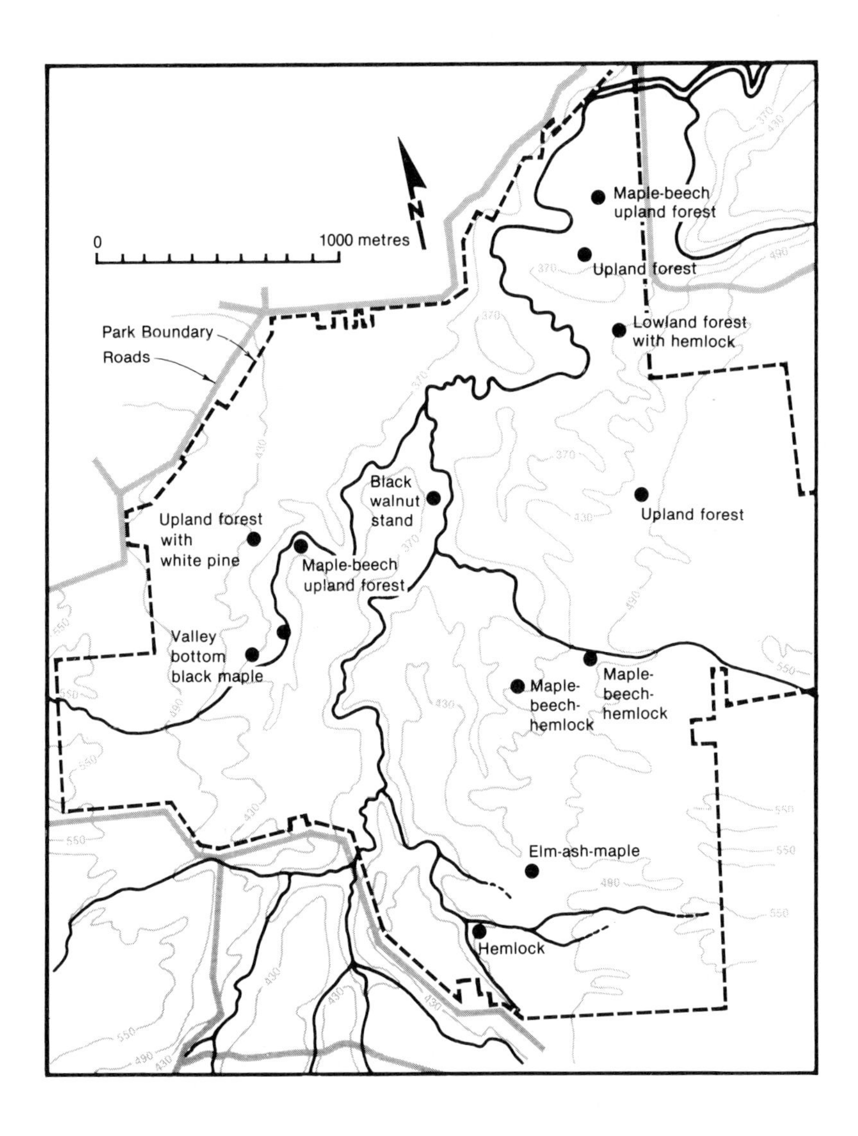

FIGURE 4,8 Short Hills Provincial Park. Individual forest stands as identified in the park planning process together with the calculated 'continuum index' for each stand (see text for meaning of values).

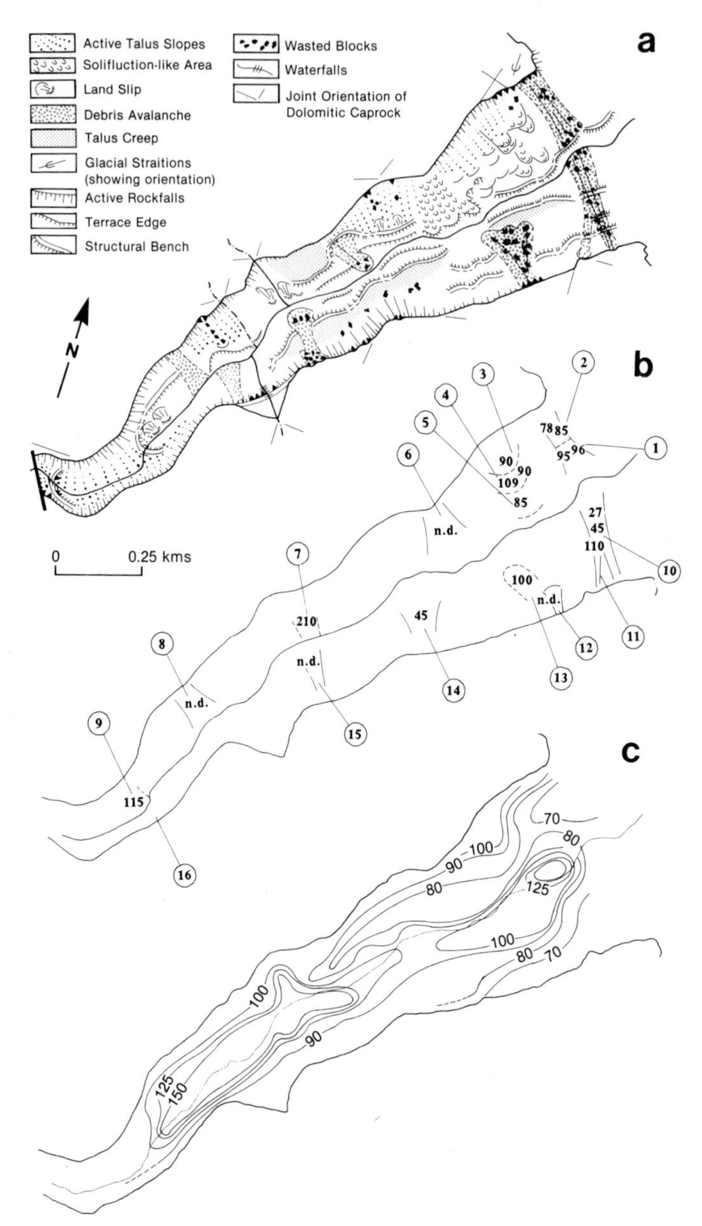

FIGURE 5,8 Forty Mile Creek, Grimsby; biophysical features: (a) the range and extent of major surficial features and events (landslips, rockfalls, talus cones etc.), (b) dates calculated by tree ring analysis for length of time, from 1975, of the last major movement of such features of sufficient impact to cause destruction of the forest cover, (c) isoline map indicating the age of forest cover on the valley sides in locations affected by only minor mass movements.

STRESS ← SIZE DISRUPTION / MAGNITUDE OF PROCESSES

SIZE (METRES)		10^2–10^4				10^1–10^2	10^1
GEOMORPHIC PROCESSES		DEBRIS SLIDES				SLUMP SOLIFLUCTION	CREEP SLOPEWASH
VEGETATIVE PROCESSES		SECONDARY SUCCESSION				GAP PHASE LARGE AREA	GAP-PHASE SMALL AREA INDIV. TREE REPLACEMENT
VEG'N COVER	TIME (YEAR)						
LOW-MID DENSE	1–2	OPEN	CEDAR		BIRCH-POPLAR	CEDAR-ASPEN POPLAR	MAPLE OAK BIRCH DEPENDS ON SEEDLINGS AT SITE
	30–40	OPEN	CEDAR	BIRCH POPLAR CEDAR	BIRCH POPLAR	BIRCH WHITE ASH MAPLE	
CLOSED CANOPY	?	OPEN	CEDAR FEW POPLAR BIRCH	BIRCH ASPEN	BIRCH POPLAR CEDAR		
STEADY STATE	?			?			
	?	MAPLE-BASSWOOD-HEMLOCK			(Based on regional classification and Ontario Land Survey)		
SLOPE UNIT		B,C	C	C(B)	B,C	B(A)	A,D
SLOPE ANGLE		25–45	30–40	20–35	15–25	10–20	0–20

FIGURE 6,8 Collingwood: model showing the complexity of interaction between geomorphic events and vegetative features (from Milne, 1982[19]).

slopes, and major gullies. Investigations show a series of complex interactions between recurrence of gully activity, stand age and forest succession. It seems highly likely that recurrence of these events, which may be related to years of higher than average precipitation, could have an impact on the tourist facilities. As gullies become very evident features near tourist sites, a reduction in landscape value could have both a scenic and an economic impact on the value of this provincially important natural feature.

Consequently, it is essential that data collection in landscape assessment studies must involve a recognition and understanding of landscape processes. The time-frame for planning should incorporate the recurrence cycle of more active biophysical events that can be shown to occur on areas of high scenic value.

SUMMARY AND CONCLUSIONS

The conclusions drawn from the discussion in the preceding sections are that traditional methods of landscape assessment may identify areas of high scenic or landscape value, but provide little relevant information for environmental planning over a longer time period. Landscape processes may significantly alter those critical environmental components which define the highest aesthetic value of an area. Furthermore, environmental or landscape changes related to these events take place within a normal planning time scale of a few decades; that is, on a biological or biophysical timescale rather than a geological or geomorphological timescale.

It has also been pointed out that, while the opportunity exists for the incorporation of landscape or scenic assessment data in a range of environmental planning scenarios, the demand for this information is poor and ill-defined. This begs the question as to the degree to which existing procedures have the capacity to incorporate, or relate to, information on likely environmental change and impact. Such environmental changes are both naturally occurring events and may also be the outcome of any conservation or management strategies developed. Hence, it is either in the context of, or in close relationship to, environmental impact assessment procedures that landscape assessment procedures need to be developed. However, such impact assessments are themselves limited in their ability to predict, being based primarily on inventory data, and thus suffer from a common set of limitations. Consequently, in many areas of resource management, improvements in technique and methodology are required. The direction suggested here is further development of landscape process analysis studies and the incorporation of this information in assessment procedures. Few examples of this approach can be cited, although possible lines of development have been suggested in the use of ecological and environmental process data in environmental impact assessment[20] and in carrying capacity determination[21] — the two areas identified by Mitchell[22] as being of concern in landscape aesthetic assessment.

REFERENCES

1. MITCHELL, B., *Geography and Resource Analysis*. London: Longman, 1979, p. 145.

2. MOSS, M.R., "Landscape synthesis, landscape processes and land classification: some theoretical and methodological issues", *GeoJournal* 7, 2, 1983, pp. 145-153 (see Figure 6,8).

3. MOSS, M.R., "Biophysical land classification schemes: a review of their relevance and applicability to agricultural development in the humid tropics", *Journal of Environmental Management* 3, 4, 1975, pp. 287-307; also GARDINER, V. and GREGORY, K.J., "Progress in portraying the physical landscape", *Progress in Physical Geography* 1, 1, 1977, pp. 1-22.

4. FEDERAL ENVIRONMENTAL ASSESSMENT REVIEW OFFICE, *Guide for Environmental Screening*, Ottawa: Federal Activities Branch, Environmental Protection Service and Federal Environmental Assessment Review Office, 1978.

5. DEARDEN, P., "Forest harvesting and landscape assessment techniques in British Columbia, Canada", *Landscape Planning* 10, pp. 239-253.

6. DEARDEN, P. and BANNERMAN, S., "Visual resource analysis in British Columbia: Current status and probable future", in *Geographical Research in the 1980s: The Kamloops Papers*. Vancouver: Tantalus Press, 1985, pp. 59-76.

7. PUBLIC WORKS, CANADA, *Proposed Improvements to the Trans-Canada Highway in Banff National Park, Km. 13 to Sunshine Road*. (Environmental Impact Statement, Vol. 1) Ottawa: Public Works, 1981.

8. VAUGHAN, A.V., "A visual analysis system", *Geographical Inter-University Resource Management Seminars*, 4, 1973-74, pp. 41-46.

9. STRAW, A., "Perceptions of the Niagara Escarpment of Southern Ontario", in ATKINSON, K. and MCDONALD, A. (eds.),

Planning and the Physical Environment in Canada, University of Leeds, Regional Canadian Studies Centre, 1985, pp. 1-22.

10. MOSS, M.R. and NICKLING, W.G., "Landscape evaluation in environmental assessment and land use planning", *Environmental Management* 4, 1, 1980, pp. 57-72.

11. LINTON, D.L., "The assessment of scenery as a natural resource", *Scottish Geographical Magazine* 84, 3, 1968, pp. 219-238.

12. NEWKIRK, R.T. *et al.*, "Quantifying landscape analysis", *Geographical Inter-University Resource Management Seminars* 5, 1974, pp. 149-163.

13. LEOPOLD, L.B., *Quantitative Comparison of Some Aesthetic Factors Among Rivers*. U.S. Geological Survey, Circular 620, 1969.

14. See MOSS, M.R. and NICKLING, W.G., 1980, *op. cit.*, Figures 2,8 and 3,8, cf. 4,8 and 5,8.

15. WIKEN, E., "Rationale and methods of ecological land surveys: an overview of Canadian approaches", in TAYLOR, D.G. (ed.), *Land/Wildlife Integration*, Ottawa: Environment Canada, Lands Directorate, Ecological Land Classification Series No. 11, 1980.

16. MOSS, M.R., 1983, *op. cit.*

17. MUELLER-DOMBOIS, D. and ELLENBERG, H., *Aims and Methods of Vegetation Ecology*. New York: John Wiley and Sons, 1974, p. 276.

18. MOSS, M.R. and ROSENFELD, C., "Morphology, mass wasting and forest ecology of a post-glacial re-entrant valley in the Niagara Escarpment", *Geografiska Annaler*, Series A, 60, 3/4, 1978, pp. 161-174.

19. MILNE, R.J., "Forest dynamics and geomorphic processes on the Niagara Escarpment, Collingwood", Unpublished M.Sc. Thesis, University of Guelph, 1982.

20. BEANLANDS, G.E. and DUINKER, P.N., *An Ecological Framework for Environmental Impact Assessment in Canada*, Halifax: Dalhousie University, Institute for Fesource and Environmental Studies, 1983.

21. MOSS, M.R., “Land processes and land classification”, *Journal of Environmental Management* 20, 1985, pp. 295-319.

22. MITCHELL, B., 1979, *op. cit.*

Water has a major influence on landscape quality whether in the form of a rushing mountain stream (**PLATE 17**) or a man-controlled impoundment (**PLATE 18**) ►

9 ON THE PERSISTENCE OF ERROR IN LANDSCAPE AESTHETICS

Louis Hamill

INTRODUCTION

Kuhn[1] and some other science theorists give the impression that modern science reacts correctly to critical analysis, and that errors of concept and method are systematically discarded, as a consequence of informed scientific communication. However, many erroneous paradigms have persisted for long periods of time, and have achieved academic respectability. In some cases, published errors define the frontiers of research. This suggests that there must be good reasons why people produce errors, why errors survive review prior to publication, why they are used in teaching, and why they often are referenced favorably in subsequent publications.

This paper identifies some factors that are related to the persistence of error in the literature of landscape aesthetics, using the Leopold method as the major example. This method, which claims to quantify the aesthetics of rivers, was published by Luna B. Leopold, a specialist in water and rivers studies.[2] Several interesting variants have been developed, not all of which have been published. A detailed analysis[3] of the Leopold method concluded that the method exhibits serious errors of concept and method, and that these errors make it unsuitable for its intended use. The demonstration of these errors has been published and the existence of this analysis is widely known. The details of the analysis have not been contested. However, the details of the critique appear not to be known by most of the people who reference it and the fact that the method contains serious errors has been widely ignored in the scholarly literature of landscape analysis. The method is still being used and recommended for use.

A TYPOLOGY OF ERRORS IN LANDSCAPE AESTHETIC RESEARCH

There appears to be an increasing incidence of errors in the literature of landscape aesthetics, not the decrease one might expect as the result of increasing knowledge. Observed examples of seven types of errors are discussed below.

Incorrect use of numbers (spurious numbers) derived from place in a classification

This problem occurs in the Leopold method and its variants. Leopold uses a list of 46 environmental variables, each of which is classified by means of a five part scheme. In order to achieve objectivity, Leopold did not use human preferences or utility for human purposes in ordering each classification. Many of the problems of the Leopold method arise from this unusual classification practice. Most subsequent variants on the Leopold method have used classification assignments which reflect human preferences or utility for human purposes. For example, the author's test of Leopold's findings started by reordering his classifications to reflect utility for human purposes.[4]

There are several important objections to the use, as statistical data, of numbers that refer to place in a classification, even when the variables classified can be measured. The classification practices used by Leopold, and by authors of variants, have demonstrated that each of the classifications used can be ordered in more than one way. If numbers can be assigned arbitrarily to one of five classes, what meaning can be assigned to the number?

Leopold, and others who have developed variants on his methods, have treated numbers derived from place in a classification as if they were cardinal numbers. In most cases, this assumption is incorrect. In some cases, the sequence of classes represent rankings of a variable, and ordinal numbers may be used. But ordinal numbers cannot be used in mathematical operations as if they were cardinal numbers.

In order to be used in ordinary arithmetic and mathematical calculations, numbers must have known intervals, usually equal. Many of the classifications of numerical variables used in the Leopold method do not exhibit equal class intervals. For example, variable 2, "Depth in Feet at Low Flow", has the following class limits: (1) under 0.5; (2) 0.5 to 1; (3) 1 to 2; (4) 2 to 5; (5) more than 5. Examination of the classification definitions used by Leopold indicates that most of those that represent numbers derived by measurement would have to be reordered in order to exhibit equal class intervals.

Incorrect use of numbers to stand for words

There are many examples of this problem in the Leopold method and its variants. In the Leopold method, for example, variable 6, "River Pattern", uses the following classification: (1) torrent; (2) pool and riffle; (3) without riffles; (4) meander; (5) braided. In this case, class labels look like numbers, but are not; they are spurious numbers and cannot be used correctly even in addition.

Use of spurious numbers in simple mathematical operations

There are many examples of two errors of this type in the Leopold method and its variants. The first error is to use spurious numbers, which represent words, in addition, subtraction, multiplication, and division. The other is to mix ordinal and cardinal numbers (and spurious numbers), and to treat them as if all were cardinal numbers. Only cardinal numbers can be used in ordinary arithmetic and mathematical operations. To mix together cardinal, ordinal, and spurious numbers is not only incorrect, but the sums of such mixtures have no useful meaning.

Use of bad data in complex mathematical and statistical operations

There are many examples of this problem in the recent literature of quantitative recreation research. A paper by Juurand, Guzelimian, and Beaman[5] illustrates the problem. The authors used data that had been collected for another purpose to produce a mathematical model of 'quality of wild rivers,' using statistical techniques which included analysis of variance and simple linear regression. Like most of the papers in Volume 2 of the *Canadian Outdoor Recreation Demand Study* report,[6] this is an impressive demonstration of statistical technique. However, the data used was taken from field cards used to collect data as part of canoe trip studies on selected rivers. Descriptions of this study, including instructions for data collection, indicate that the data are not good enough to use in this complex statistical analysis. The first problem is that great accuracy was not sought in the collection of data. For example, "Flow Pattern" was classified as: 1, high; 2, medium; 3, low. "Predominant Fluvial Process" was classified as: 1, erosion; 2, erosion and deposition; 3, deposition. Many of the other classifications used were similarly uninformative and of low accuracy. More important, the data collection format used a variant of the Leopold method, in which some classifications reflected human preferences and some did not; all of the numerical defects of the Leopold classification system, discussed above and in earlier papers are found in the data used in this study.[7] Yet this study exhibits the most complex manipulation of data produced of any variant on the Leopold method. For reasons explained above, much of the data used in this study cannot be used correctly even for simple arithmetic calculations; it is clearly unsuited to the complex procedures actually used.

Use of data which does not satisfy requirements of the model

All statistical operations, including calculation of a mean, have definite requirements, which are available in the literature. The requirements of standard procedures may be found in textbooks, but one may have to search articles and specialist books and reports to identify the requirements of less common procedures. Generally, the more complex the technique, the more demanding is the model. Personal experience with much of the literature reporting findings of statistical studies indicates that it is common practice to give lip service to the demands of the model, and then to ignore many of the details. The report discussed in the preceding section illustrates this problem. Indeed, many of the studies published in the report of the *Canadian Outdoor Recreation Demand Study* are examples of elaborate procedures using poor data.

Use of numbers to support, derive, or demonstrate meaningless, spurious, or useless concepts

This problem is evident in the Leopold method, in the concept 'uniqueness ratio,' which was described by Leopold[8] as follows: "If a site factor is, for example, one among 12 of the same category, the site shares this characteristic with 11 others. It is unique in the ratio of 1 to 12 or its uniqueness ratio is 1:12 (.08)." This concept is hard to comprehend when it is applied to one environmental variable, such as river pattern. But it becomes meaningless when uniqueness ratios are added and/or averaged. That is, the meaning cannot be stated adequately in English, the language used in Leopold's article. The concept is also useless, for the same reason; if the meaning cannot be stated, it cannot be known, and therefore serves no human purpose (except perhaps to entertain). This problem arose because of Leopold's initial decision to classify resources 'objectively,' by avoiding the use of classification criteria that reflected human preferences.

Use of concepts without adequate operational definition

Leopold uses 46 environmental factors in the checklist which forms the basis for his analysis. Many of the concepts used could be measured in different ways by people who did not have access to the definitions and/or procedures used by Leopold in applying the method. "River Fauna" is one of many variables for which lack of operational definition could cause problems: how does one determine which of five classes to use in the range of conditions from: (1) none to (5) large variety? This problem can be solved only by providing precise instructions to potential users of the method. The use of imprecise instructions, or no instructions, instead of valid operational definitions, is one cause of the poor data produced in the Leopold-type inventory in the Canadian Wild Rivers Survey, which data were used in the study by Juurand, Guzelimian, and Beaman, discussed above.

FACTORS ACCOUNTING FOR THE PERSISTENCE OF ERROR

Many cultural and institutional factors may account for the creation and persistence of error in the scholarly literature and landscape aesthetics. A complete consideration of these factors is beyond the scope of this article. The factors discussed briefly below seem particularly relevant to understanding the creation and persistence of the kinds of errors summarized above.

Hierarchies of Reputation, Position and Power

The existence of hierarchies of reputation, position and power in the academic/scientific worlds is evident and needs no demonstration. Persons at upper levels of hierarchies are almost beyond serious challenge, for a number of reasons. For example, the excellent reputation of L.B. Leopold for his work on rivers and related subjects contributes to unwillingness to accept criticism of his work on landscape aesthetics. A more important factor is that people who are in upper levels of hierarchies usually wield real power, and this is as evident in the academic world as in others. Many people seek to avoid making powerful enemies who may affect their career advancement negatively.

Academic Beliefs

All academic disciplines, at any particular time and place, exhibit a widespread belief in particular concepts, approaches, methods, research and teaching styles, and the like. Two such beliefs that have affected the acceptance of certain kinds of work, and the persistence of errors, are: (1) belief in the possibility of attaining objective knowledge of phenomena;

and (2) belief in the superiority of quantitative knowledge. The preceding discussion of specific errors in aesthetic analysis makes reference to cases in which belief in these concepts either led to errors or to the persistence of errors, or both.

Evidence of Affiliation to an Established Position or Group

There is organized support for each of the major approaches in aesthetic analysis of landscapes, as in other geographic subjects. A major requirement for a positive judgment of academic merit in a work usually is evidence of support for one of the dominant approaches. If a new or innovative approach is proposed, there must be evidence of adherence to basic philosophical positions, such as belief in the search for objective knowledge and the importance of quantification. In other words, evidence of acceptable ideological affiliation in an identifiable scholarly discipline is more important than care in avoiding errors of procedure, fact, or conceptual detail.

Novelty and Innovation

The requirement to publish frequently has forced many academics to seek subjects which have a high probability of being published. Evaluators of academic merit place a high value on novel or innovative approaches, concepts, techniques and on subjects that are conceptually or technically interesting, or otherwise unusual. Complex statistical or mathematical models, and new uses of such models, appear to be treated automatically as being new and innovative. In addition, what is seen as new or innovative often is evaluated leniently. These practices encourage the use of complex concepts and techniques, which account for much of the recent work in landscape aesthetics, and may explain the publication of some of the specific errors listed earlier.

Problems arising from affiliation with an established position and emphasis on novelty and innovation are illustrated in the discussion of the Leopold method in Mitchell's, *Geography and Resource Analysis.*[9] This book seems to have been motivated by the belief that recent work in geography and related subjects, especially new concepts and quantitative approaches, have great relevance to understanding of natural resources, environment, and the like. As a consequence, there is little critical analysis of concepts and approaches that are identified with the "new geography."

Although the author's critique of the Leopold method indicated that the uniqueness ratio is meaningless when applied to more than one phenomenon at a time, Mitchell's discussion implies a rejection of this finding. However, he ends his discussion of the method with a summary of five questions which he says were raised by Leopold, and which raise doubts about the validity and utility of the uniqueness ratio concepts, but in such a way as to leave the validity of these concepts intact. He writes:

> Although these points represent potential weaknesses in Leopold's sytem, they also indicate future research directions. Those who are critical would overcome such difficulties.[10]

Mitchell's long review supports the continued use of the Leopold method, with the proviso that some slight changes may be made in details of the method, following additional research with this objective. But he provides no evidence to support the belief that additional research can overcome the fundamental errors in the uniqueness ratio concept, either when it is applied to spurious numbers, or when its use produces spurious numbers. Mitchell simply ignores the evidence that such fundamental errors may exist.

Peer Review

Research proposals and the products of scholarly work typically are subjected to some form of peer review, usually involving only a small number of reviewers; sometimes only one reviewer is used. Reliance on peer review for evaluating research assumes that reviewers are unbiased and use all relevant criteria and procedures properly and fairly in the evaluation of research proposals, research results, written reports, and other products. However, these assumptions have been subverted so widely that peer review is no longer reliable for fair, unbiased, accurate, and complete reviews of research reports and other writings. There is convincing evidence that academics have learned how to use peer review to the advantage of individuals and groups.

People having compatible interests often act in concert to use peer review for mutual advantage. There are two major strategies. One is to support persons who share similar beliefs or interests, or who share some other affinity. The other major strategy is to oppose the work of persons or groups believed to be antagonistic to the interests of a person or group. These strategies may be documented in the practices of individuals and groups in referencing published and unpublished work. In the literature of landscape analysis, the ideological, technical, or other affiliations of individuals may usually be identified with reasonable accuracy in the items included in footnotes and reference lists, as well as in the items excluded.

The exclusion, or non-recognition, of critics is as important in protecting individual and group interests as the positive support of adherents to a specific view.

Academic Merit

Academic peer review is intended mainly to identify and reward academic merit. But academic merit is not defined operationally; the evaluation of academic merit typically operates without requiring the use of specific criteria. Reviewers, who are usually academics, are expected to be able to recognize academic merit without the use of specific criteria. The lack of specific criteria sets academic evaluation apart from evaluation of publications and research proposals in applied fields, in which specific criteria, including utility or superiority for achieving specific objectives, usually are used. Peer review may be used, for example, to advance the objectives of reviewers, which may not be identical with the assumed objectives of fair and accurate evaluation.

CONCLUSIONS

Among the factors that account for the persistence of error are: hierarchies of reputation, position and power; academic beliefs, including beliefs in the possibility of objective knowledge and in the superiority of quantitative knowledge; evidence of affiliation to established positions or to affinity groups; and emphasis on the value of novelty and innovation. Institutional practices that encourage the production and persistence of error include: peer review of publications and research proposals; use of the concept 'academic merit' in evaluating written work; and lack of specific criteria for judging academic merit.

Many of the institutional practices of universities seem to be intended mainly to identify and reward ideological orthodoxy. The most fundamental practice is the judgment of academic merit by means of the process of peer review. Peer review is an effective mechanism for identifying and judging academic merit, which, in many cases, seems to be identical with ideological orthodoxy.

Since academic peer review does not adequately identify and correct errors of procedure, errors of fact, and errors of concept, it is concluded that it is not intended primarily to do so, but to meet other objectives, some of which are identified in the text of this article.

Ideological orthodoxy is now generating and perpetuating a large and increasing amount of conceptual and procedural error in the literature of aesthetic landscape analysis. If the error content of the literature of landscape aesthetics is to be reduced significantly, better criteria and procedures for reviewing research proposals and written reports is one possible solution. Improvements would include the use of specific criteria for evaluation, including criteria of utility, accuracy, operationality, and non-complexity. Such criteria would be applied to all research proposals and products of research and writing, including novel and innovative concepts and procedures. These should be judged by comparison with available concepts and procedures; novel concepts and procedures should not be evaluated leniently, as they are at present.

Procedural changes, such as the use of specific criteria of evaluation, are not likely to be instituted on a significant scale until there are fundamental chances in belief and in institutional practices. Without the necessary fundamental changes, it may be expected that error will continue to proliferate in this literature.

REFERENCES

1. KUHN, T.S., *The Structure of Scientific Revolutions*. Chicago: University of Chicago Press, 1970, 2nd edition.

2. LEOPOLD, L.B., "Landscape aesthetics: how to quantify the scenics of a river valley", *Natural History*, October 1969a; and LEOPOLD, L.B., *Quantitative Comparison of Some Aesthetic Factors Among Rivers*. Circular 620, U.S. Geological Survey, Washington, D.C., 1969

3. HAMILL, L., "Statistical Tests of Leopold's System for Quantifying Aesthetic Factors Among Rivers", *Water Resources Research*, Vol. 10, 3, 1974, pp. 395-401; and HAMILL, L., "Analysis of Leopold's Quantitative Comparisons of Landscape Aesthetics", *Journal of Leisure Research*, 7, 1, 1976, pp. 16-28.

4. HAMILL, L., *op. cit.*

5. JUURAND, P., GUZELIMIAN, V. and BEAMAN, J., "Perception of Quality of Wild Rivers", *Canadian Outdoor Recreation Demand Study*. Vol. 2, Toronto: Ontario Research Council on Leisure, 1976, pp. 204-208.

6. *Ibid.*

7. HAMILL, L., *op. cit.*

8. LEOPOLD, L.B., *op. cit.*, 1969b, p. 5.

9. MITCHELL, B., *Geography and Resource Analysis*. London and New York: Longman, 1979, p. xiii.

10. MITCHELL, B., *op. cit.*, p. 156.

Canyons of the land; Canyonlands National Park, Utah (**Plate 19**), and downtown Calgary, Alberta (**PLATE 20**). ►

PAY MORE?
CENTRE ST S
LITZ
SUN

PART III

APPLICATIONS

PLATE 21 Yosemite Valley, California. ►

10 SCENIC PERCEPTION: RESEARCH AND APPLICATION IN U.S. VISUAL MANAGEMENT SYSTEMS

Robert M. Itami

INTRODUCTION

Twenty years have passed since some of the first efforts to systematically rate the visual quality of landscapes in the United States were attempted.[1] From these earlier works has emerged a fuller understanding of the nature of human landscape perception as research and applied visual assessments continue. This work has been spurred by the pragmatic needs of resource and land use planners and policy makers to incorporlate 'environmental intangibles' into the decision making process.[2] In response to this need, landscape architects, geographers, foresters, recreation experts, and psychologists have brought their interests and expertise to the issue of measuring and assessing scenic quality, resulting in a rich array of conceptual approaches.

Recent appraisals of the literature by Zube, Sell and Taylor[3] and Daniel and Vining[4] have attempted to categorize the literature according to major paradigms of landscape appraisal. Both reviews have identified four schools of thought which have emerged in the last 15 to 20 years. The first, the expert paradigm, has generally been application-oriented and is evidenced by the visual management procedures developed by the United States Department of the Interior (USDI) Bureau of Land Management, the United States Department of Agriculture (USDA) Forest Service, and the Department of Transportation (DOT) Federal Highway Administration. This paradigm is typified by Zube *et al.* as a process of evaluation by skilled and trained observers based on principles from art and design, ecology or resource management.[5]

The second paradigm is labelled the "psychophysical" by both Zube *et al.* and Daniel and Vining. According to Daniel and Vining, psychophysical methods of scenic assessment "seek to determine mathematical relationships between the physical characteristics of the landscape and the perceptual judgements of human observers."[6] This paradigm stems from a well established branch of psychology with rigorous standards of reliability, validity and sensitivity.

The third paradigm, called the "cognitive paradigm" by Zube *et al.* and the "psychological model" by Daniel and Vining, involves "a search for human meaning associated with landscapes or landscape properties. Information is received by the human observer and, in conjunction with past experience, future expectation, and sociocultural conditioning, lends meaning to landscape."[8] Work by Kaplan and Kaplan[9] with their information processing model and Appleton's[10] prospect and refuge theory are examples of this paradigm.

The "experiential paradigm"[11] or the "phenomenological model"[12] views the interaction between human and environment as a dynamic encounter and "places emphasis on individual subjective feelings, expectations and interpretations."[13] This paradigm has had little impact on applied scenic assessment and visual management and will be discussed only in the context of future directions in the conclusion of this chapter.

This chapter is concerned with the impact of landscape perception research on applied landscape assessment in the United States. In particular, the chapter reviews current procedures institutionalized in the USDA Forest Service, the USDI Bureau of Land Management and the DOT Federal Highway Administration.

A REVIEW OF APPLIED VISUAL ASSESSMENT AND VISUAL IMPACT PROCEDURES

The USDA Forest Service,[14] the USDI Bureau of Land Management (BLM),[15] and the DOT Federal Highway Administration (FHWA),[16] have all published procedures for visual classification, scenic assessment, public sensitivity assessment and visual impact assessment. Each of these federal agencies has employed these procedures in response to the need to include visual/aesthetic values in their respective decision making processes. The USDA Forest Service visual management system and the USDI BLM visual resource management procedures are generally applied in predominantly natural landscape settings. The DOT Federal Highway Administration visual impact assessment procedures are applied in a broad range of contexts, primarily urban, and in relationship to impacts related to construction of transportation facilities. Though each vary in detail, the following basic components provide a framework for comparison.

Classification of Landscape Character

The purpose of landscape character classification is to provide a "frame of reference" for scenic quality assessments in order to avoid comparing 'apples and oranges' in the assessment process. The USDA Forest Service's Visual Management System (VMS) uses broad physiographic sections as defined by Fenneman[17] to define 'Character Types.' These may be subdivided according to local variations into sub-types.[18] The FHWA Visual Impact Assessment (VIA) procedure also use physiographic sections to define the regional context for visual impact assessment. These are subdivided into landscape units by identifying local variations in landscape pattern. By overlaying a viewshed map, which delineates the extent of views from the highway in question, Visual Assessment Units are defined.[19] The BLM Visual Resource Management (VRM) procedure[20] does not specify the delineation of landscape character units, though in the interim guidance[21] states that physiographic units are used as the context for visual assessment.

Assessment of Scenic Quality

The purpose of the assessment of scenic quality is to map the variation of scenic quality across the landscape in order to identify and protect landscapes of highest visual amenity. Though all three of the procedures in question are based on the expert paradigm of landscape perception, they differ in their criteria of assessment. The Forest Service VMS assesses scenic quality based on visual variety of landforms, rockforms, vegetation pattern, and water form within each character type or subtype to derive distinct (class A), common (class B), or minimal (class C) variety classes. The assumption is that "all landscapes have some value, but those with the most variety or diversity have the greatest potential for high scenic value."[22]

The FHWA VIA uses vividness (the visual power or memorability of the landscape components as they combine in striking and distinctive visual patterns), intactness (the visual integrity of the natural and man-built landscape and its freedom from encroaching elements), and unity (the visual coherence and compositional harmony of the landscape considered as a whole).[23]

The BLM VRM procedure rates the scale and ruggedness of landforms; vegetation variety, the presence, size dominance and movement of water; the intensity of colours; the influence of adjacent scenery; the scarcity or uniqueness of scenery; and the amount of negative influence from cultural modifications to assess scenic quality. Each of these "seven key factors"

Table 1,10 BLM SCENIC QUALITY SCORING CRITERIA

KEY FACTORS	RATING CRITERIA AND SCORE		
LANDFORM	High vertical relief as expressed in prominent cliffs, spires, or massive rock outcrops; *or* severe surface variation or highly eroded formations including major badlands or dune systems; *or* detail features dominant and exceptionally striking and intriguing such as glaciers. **5**	Steep canyons, mesas, buttes, cinder cones, and drumlins; *or* interesting erosional patterns or variety in size and shape of landforms; *or* detail features which are interesting though not dominant or exceptional. **3**	Low rolling hills, foothills, *or* flat valley bottoms, *or* few or no interesting landscape features. **1**
VEGETATION	A variety of vegetative types as expressed in interesting forms, textures, and patterns. **5**	Some variety of vegetation, but only one or two major types. **3**	Little or no variety or contrast in vegetation. **1**
WATER	Clear and clean appearing, still, or cascading white water, any of factor in the landscape. **5**	Flowing, or still, but not dominant in the landscape. **3**	Absent, or present, but not noticeable. **1**
COLOR	Rich color combinations, variety or vivid color; *or* pleasing contrasts in the soil, rock, vegetation, water or snow fields. **5**	Some intensity or variety in colors and contrast of the soil, rock, and vegetation, but not a dominant scenic element. **3**	Subtle color variations, contrast, or interest; generally mute tones. **1**

Table 1,10 (continued)

KEY FACTORS	RATING CRITERIA AND SCORE		
INFLUENCE OF ADJACENT SCENERY	Adjacent scenery greatly enhances visual quality. **5**	Adjacent scenery moderately enhances overall visual quality. **3**	Adjacent scenery has little or no influence on overall visual quality. **0**
SCARCITY	One of a kind; *or* unusually memorable, *or* very rare within region. Consistent chance for exceptional wildlife or wildflower viewing etc. **5 +**	Distinctive, though somewhat similar to others within the region. **3**	Interesting within its setting, but fairly common with the region. **1**
CULTURAL MODIFICATIONS	Modifications add favorably to visual variety. **5**	Modifications add little or no visual variety to the area. **0**	Modifications are extensive and scenic qualities are substantially reduced. **– 4**

SCENIC QUALITY: A = 19 or more; B = 12-18; C = 11 or less.

are rated from a table ranging from -4 to 6, factor ratings are then added to derive three scenic quality classes (A, B, and C). This is shown in Table 1,10.

The Assessment of Public Sensitivity

Sensitivity levels "measure people's concern for the scenic quality."[24] The Forest Service first categorizes use areas such as roads and recreation sites according to their relative national or local importance in terms of use volumes, use duration and size. From these areas viewsheds are mapped to derive three sensitivity levels (1, 2 and 3). Each of these levels is subdivided by distance from the use area into foreground, middle ground and background distance zones to create a composite sensitivity level map. In cases where two or more levels overlap, the more restrictive is assigned.

The FHWA procedure is less explicit about mapping public sensitivity because of the more complex, often urban context of the landscape. Their recommendations include considering the number of users, the perceived compatibility of the project to its surroundings, the type of activity (such as outdoor recreation which may alter the awareness of the viewer to his or her surroundings) and the local cultural values and concerns of the community. This information may be recorded graphically, in a narrative, or with checklists or matrices.

The BLM procedure, like the Forest Service system, first identifies three sensitivity levels based on user attitudes to the project and the use volumes. Once the sensitivity levels are determined and mapped, distance zones from key use areas mapped into foreground/middleground, background and seldom seen zones. As in the Forest Service procedure, where there may be an overlap of distance zones, the more restrictive zone is selected.

Management Objectives

Visual Quality Objectives (USFS),[25] or Management Classes (BLM)[26] prescribe the degree of visual alteration which is acceptable in a given landscape. In both the USFS and BLM procedures, management objectives are assigned based on specific combinations of scenic quality, public sensitivity and distance zones. Management objectives range from most restrictive (preservation for USFS or Class 1-ecological changes only for BLM) to maximum modification and rehabilitation at the least restrictive level. The FHWA does not have a comparable step in its procedure.

Visual Impact Assessment

Visual impact assessment procedures are undertaken to predict the change of character and quality of the visual landscape from proposed developments. Procedures vary greatly in this aspect of visual management. The USFS does not include a procedure for Visual Impact Assessment in its VMS, however, on specific projects, variations on a procedure called 'visual absorption capability' (VAC) are employed. According to Anderson *et al.*,[27] VAC is "a tool to assess a landscape's susceptibility to visual change caused by man's activities." In a typical VAC study biophysical variables such as slope, vegetation pattern, soil colour contrast and site recoverability are used to estimate a site's ability to screen or camouflage proposed activities. It is a generic approach to a wide range of visual impacts and therefore does not accurately predict the impact of specific project proposals. VAC does not explicitly measure changes in scenic quality or visual character. Other techniques used by the forest service include various computer-graphic applications to simulate changes.

The FHWA VIA uses a combination of narrative descriptions and favours the use of photographic simulation over artists conceptions to depict project proposals. View response is inferred based on existing visual quality and professional judgements of the scale and compatibility of the project to the existing landscape, or viewers may be asked to rate visual impacts from photographic simulations of projects. No set procedure for either approach is detailed.

The BLM VRM system is the most well-documented and systematic of the visual impact procedures. First the existing landscape is characterized according to the form, line, colour and texture of the land/water surface, vegetation, and structures. This provides baseline information on existing visual conditions. Next the proposed activity is characterized for its form, line, colour and texture characteristics. Each element is rated for its contrast (strong, moderate, weak or none) to the existing landscape. The result is allocated to one of three classes of impact: a) contrast can be seen but does not attract attention, b) attracts attention and begins to dominate, c) demands attention and will not be overlooked by the average observer. There is no explicit rating of the change in scenic quality but the underlying assumption is that as contrast ratings increase, scenic quality decreases.

The visual classification, assessment and impact procedures described above have been developed with the intent of making systematic and explicit decisions regarding landscape scenery by articulating the criteria and judgements used to assess scenic quality and visual impacts. Table 2,10 shows a comparison of the three approaches. The development of these techniques has played an important role for over a decade in providing a

Table 2,10 COMPARISON OF U.S. VISUAL ASSESSMENT AND IMPACT SYSTEMS

	USDA FOREST SERVICE	USDI BUREAU OF LAND MANAGEMENT	DOT FEDERAL HIGHWAY ADMINISTRATION
LANDSCAPE CLASSIFICATION	Physiographic Units Character Types, Subtypes	Physiographic Units set landscape context	Physiographic Units Viewsheds from Highway Corridors
SCENIC ASSESSMENT	Visual Variety of Landforms, Waterforms, Rockforms and Vegetation	Landform scale & Ruggedness Vegetation Variety Water size/dominance Color Intensity Adjacent Scenery Uniqueness Negative Cultural Features	Vividness Intactness Unity
PUBLIC SENSITIVITY	Importance of Use Areas Distance Zones Viewsheds	User attitudes and use volumes Distance Zones Viewsheds	Use Volumes Landuse compatibility Local Cultural Values
MANAGEMENT OBJECTIVES	Scenic Assessment and Public Sensitivity combined to generate management objectives	Scenic Assessment and Public Sensitivity combined to generate management objectives	Not Applicable
VISUAL IMPACT ASSESSMENT	Visual Absorption Capability, estimates site screening potential	Visual Contrast rating for existing landscape and proposed uses.	Photographic Simulation

means of incorporating landscape values into the planning and management process. The pragmatic bent of the "expert paradigm" represented by the procedures described here have resulted in an impressive number of studies with consequences for literally millions of acres of public and private land. Yet there has been little evidence that these procedures have benefited from the growing body of environmental perception research. Each step in the procedures described here is based on professional judgement and may be questioned for their reliability and validity.[28] The next section will summarize landscape research that has most relevance to applied visual assessment and management techniques.

IF IT WORKS — DON'T FIX IT!

Laughlin,[29] studied the attitudes of landscape architects in the USDA Forest Service toward the visual management system. In a survey of 260 in-service landscape architects using the visual management system, Laughlin received an impressive ninety-five percent response rate (236 respondents). The survey questionnaire sought to determine: a) the landscape architects' views of the system as an whole; b) the landscape architects' views of the primary elements that comprise the system; and c) the landscape architects' perceptions of the attitudes of other managers in the Forest Service toward the system. Laughlin found:

> A very high number of landscape architects in the USDA Forest Service believe that the Visual Management System: has succeeded in quantifying the visual resource (90 percent); has changed the way the visual resource is managed on the ground (95 percent); has given the visual resource added weight in multiple-use decision-making (96 percent); and is not a passing fad (96 percent). The data indicate that the Visual Management System has been accepted by the landscape architects as a means of managing the visual resource, and is making a positive difference in the treatment of the visual resource within the Forest Service.[30]

In regard to the elements of the VMS system, Laughlin found eighty-four percent of the landscape architects felt that Variety classes are accurate indications of the scenic quality of an area and that the VMS provides an accurate measurement of scenic quality (62 percent).[31] However almost half of the respondents did not feel Sensitivity Levels are accurate measurements of the public concern for scenic quality.[32]

Grden[34] conducted a similar study comparing the BLM and Forest Service systems. He found the same problems Laughlin identified in that sensitivity levels in both the Forest Service VMS and the BLM VRM were difficult to understand and apply. Grden's analyses of the scenic quality procedures are interesting to note. Forest Service personnel found the vague A-B-C categories harder to apply than the BLM personnel found the numerical rating system. However, more BLM personnel felt the BLM scenic quality procedure was biased, compared to the Forest Service personnel view of their own scenic quality procedure. In other words, quantitative techniques were easier to apply but the more complex procedures (i.e. more variables) were thought to be more biased. This may help explain the fairly high satisfaction Forest Service landscape architects have for the scenic assessment procedure.[35] The less explicit procedure of the VMS scenic quality method is more flexible in interpretation and possibly demands more professional judgement, allowing the landscape architect more leeway in matching the variety classes closer to his or her own intuitive judgement of scenic quality. This gives the landscape architect more confidence in the results. This is speculative but makes sense within the context of the expert paradigm which states that the trained landscape architect *is* the best judge of scenic quality.

This opens discussion to a particular concern of researchers who, using rigorous measures of validity and reliability, are less satisfied with the visual management system than are its practitioners.

SETTING STANDARDS: THE VIEW FROM THE IVORY TOWER

Craik and Feimer have called for technical standards for visual assessment procedures similar to those set by the American Psychological Association. They state:

> the time has come to address the question of establishing technical standards for Visual Resource and Impact Assessment (VRIA) systems. In making an informed decision to adopt a VRIA system or to put into operational use a VRIA system developed in-house, agency staff must know what information it is appropriate to seek and require regarding performance characteristics and psychometric properties and must apply suitable criteria of psychological measurement in appraising VRIA systems.
>
> ..... Clearly the next step is to expect that agency-level manuals will move beyond mere description of procedures and forms to include psychometric findings on reliability, validity, generalizability and utility.[36]

According to Smardon:

> Reliability refers to the consistency and precision of measurement; it reflects the degree to which the obtained measures are replicable in the same or highly similar circumstances, as well as the attainable level of discrimination among the objects of interest. In the context of VIA, reliability represents the degree to which a measure accurately reflects variations among landscape and land use conditions. Validity provides an estimate of the degree to which a method is able to capture meaningful variations in the aesthetic quality of the landscape and to depict the impact of land use activities upon it. It should also be noted that reliability has important implications for validity in that the reliability of a measure delimits its attainable validity.[37]

Daniel and Vining define generality in the context of utility. They state:

> Utility of a method is usually gauged in terms of efficiency and generality. Efficient methods provide precise, reliable measures with relatively low costs in time, materials and equipment and personnel. Generality refers to the extent to which a method can be applied successfully, with minor modifications to a wide range of landscape-quality assessment problems. Utility is of obvious importance for methods that have been developed in practical, applied context.[38]

In the face of these standards how well do applied visual quality and visual impact assessments stand up?

UNDER THE MICROSCOPE — APPLIED SCENIC ASSESSMENT AND VISUAL IMPACT

Studies Under Scrutiny

There has been little empirical research on the reliability, validity, generalizability, or efficiency of the Forest Service VMS or FHWA VIA. Most recent studies have concentrated on the BLM VRM system. In the following discussion, studies specific to the BLM system are presented first followed by research studies which have implications for a broad range of studies and approaches.

Kopka and Ross[39] studied the reliability of trained judges in assessing the seven BLM scenic quality variables and the three distance zone variables.

They found with four judges, only one of the BLM variables (variety) could be judged with acceptable standards of reliability across the four judges. Miller[40] tested the validity of the BLM scenic quality rating procedures. Fifty-eight subjects were asked to rate 40 colour slides for their overall preference for those scenes on a one to five scale. Second, the subjects were given training on the BLM VRM program and procedures for rating the BLM scenic variables. The subjects were asked to rate a subset of 20 slides from the original 40 for the BLM scenic quality variables (except water which was not represented by any slides). The scores for each slide were weighted and summed according to the BLM procedure and transformed to a scale with a maximum of five for comparision with the preference ratings. Miller found that the BLM scenic quality score did indeed reflect landscape preference for the same scenes. In addition, he found, using a paired t-test, that there was no significant difference in mean BLM scores between landscape architects in the respondent group (N = 26) and subjects in other resource management fields (N = 32). Miller, using a content identifying methodology, found that BLM variables, other than landform, do not "provide an understanding of causal factors affecting landscape preference."[41]

In one of the most thorough studies of visual impact assessment methods, Smardon *et al.*[42] tested a wide range of landscape dimensions for their predictive power as indicators of visual impact (see also Craik and Feimer; Feimer *et al.*[43]). The thirteen dimensions selected included: ambiguity, colour, compatibility, complexity, congruity, form, intactness, line, novelty, scenic beauty, texture, unity and vividness. The impact evaluation descriptors were importance (of an element) and severity (of visual impact). Scenic beauty as defined by Daniel and Boster[44] was used as a general index of visual quality. Photographs representing a wide range of physiographic landscape types and land use activities were selected as stimuli for the study. For each landscape represented, pre- and post-impacts for each site were simulated with varying land use activities to provide comparisons of scenic beauty changes resulting from land use impacts on the same site. Smardon and associates found single rater reliabilities for all dimensions to be well below accepted standards (.7 is acceptable, Smardon *et al.* found reliability coefficients averaging only .26 and .21 for before and after impact assessments).

Out of the thirteen landscape dimensions only four (compatibility, congruity, intactness and form) were significantly correlated with change in scenic beauty. Subjects were also asked to list and rank criteria they used in rating scenes for scenic quality. Smardon and his associates (pp. 94-95) found the most "often mentioned as aesthetic factors related to severity of visual impact were naturalness, fittingness, compatibility and appropriateness of the intrusion. The most prominent physical criteria cited were change in colour and form qualities and magnitude of the intrusion."[45]

From these findings, Smardon's team proceeded to develop a prototypical visual impact assessment system which was then tested against the same standards for reliability and validity. This test showed the difficulty and complexity of visual impact systems. With the new manuals and careful training, reliability levels still fell below acceptable levels. In the test of validity, ratings from the new procedure were correlated with change in scenic beauty ratings. The results showed:

> First, it is much easier for people to judge the visual impact of structures than landform/water bodies or vegetation. Second, the variables that most consistently behave similarly to changes in scenic beauty are scale contrast, spatial dominance, for all situations; and texture, form, line and colour contrast for structures only (Smardon *et al.*).[46]

STONES AND GLASS HOUSES

The proponents of the psychophysical paradigm have presented fairly strong research findings which raise some serious questions about currently applied visual resource management systems. Is it possible to design a generalized method which will meet these standards? Smardon *et al.*[47] have shown the difficulty of meeting standards even in procedures carefully designed from research findings. Are the psychophysicists throwing stones from glass houses.

On-going applied psychophysical research in forest scenic beauty can be found in the work of Daniel and Boster[48] and the scenic beauty estimation model (SBE). This work has concentrated on near views of forest stands to predict the visual impact of various forest management practices.[49] Using standard forest mensuration measures of variables such as amount of downed wood, tree diameter, tree density, crown-cover canopy, height of understory vegetation, and others, coefficients are calculated for each forest from a step-wise multiple regression against scenic beauty estimates. SBE's are an interval scale of scenic beauty ratings based on careful selection of photographs and public perceptions of scenic quality which are transformed based on psychophysical scaling techniques and signal detection theory.[50] The results allow predictions of changes in scenic beauty by calculating the changes in the physical variables from pre-harvest to post-harvest conditions and recalculating the SBE's by using the coefficients for the variables for that stand.

This method not only meets the technical standards for reliability and validity but also is an efficient method, since the same variables which are used to calculate SBE's are also used by timber management. The drawback lies in the generalizability of the method. Brown and Daniel[51] developed SBE models for Ponderosa Pine forest in three different forests (two in Arizona, one in Colorado). Even though the three forests were all of the same forest type (Ponderosa Pine), new coefficients had to be calculated for each forest (though the variables remained the same). In addition, SBE's were calculated from responses to near-view (foreground) photographs of the forest stand. These findings raise some important questions about the generalizability of the method. Can medium and long views be predicted from the same variables? Is it possible to use the same methods on a broad range of landscape contexts from urban to rural?

At least some of these questions have been answered by researchers who have applied the SBE method in urban landscapes. Im[52] and Anderson and Schroeder[53] both found the SBE method a valid and reliable measurement technique in urban contexts, but, as can be expected, different physical variables explained the scenic beauty measures for urban scenes as compared to forest settings. If this pattern holds true, then every landscape must be considered unique, therefore, testing and retesting of each landscape context for public preferences, physical variables and their coefficients must be calculated. Each landscape context comes with its own physical makeup and therefore no uniformity of variables can be designated between contexts. This seems to support the contention mentioned in applied visual management techniques that one cannot compare the apples and oranges of landscape types.

This dilemma is also pointed out by Daniel and Vining who conclude of the psychophysical approach:

> High levels of precision and consistency have been achieved, but to some extent this has been at the expense of generality; psychophysical models are typically very specific and are restricted to a particular landscape type and to a specified viewer population and perspective. Still, for those landscape quality assessment contexts where psychophysical models have been developed, no other approach has come so close to meeting the criteria of the ideal assessment system. Certainly no other objective perceptual approach relates landscape quality so systematically to the objective properties of the environment. The psychophysical methods may be weakest on the psychological side — human response is typically restricted to a single human dimension.[54]

ASKING THE HARD QUESTIONS

The previous discussion has concentrated on the existing status of applied visual management systems of the Forest Service, BLM and FHWA.

Criticisms coming from researchers largely from the psychophysical school have been reviewed to illuminate some of the difficulties in achieving standards of reliability, validity and generalizability. It is apparent that the expert paradigm and the psychophysical paradigm both have strengths and weaknesses. The expert paradigm sacrifices validity and reliability for the cause of utility — getting the job done. The psychophysical school tends to sacrifice generalizability and utility to achieve high standards for reliability and validity. This dilemma is apparent in the work of Smardon *et al.*[55] in their attempts to marry the two approaches. These efforts were not successful; however, they are to date, one of the best concentrated efforts to improve existing visual management procedures.

Parallel to this work is a growing body of research in the cognitive paradigm. While the psychophysical and expert paradigm select landscape variables based on assumed relationships between perceived scenic quality and the physical landscape, the cognitive paradigm attempts to provide a theoretical explanation of why people have visual preferences for different landscapes. Two examples of this paradigm which most directly address the problem of applied scenic assessment have been Appleton's[56] Prospect and Refuge Theory and Kaplan and Kaplan's[57] information processing model. This discussion will focus on the Kaplans' information processing model which has been applied in several different visual assessment studies.[58]

The Kaplans' use a "content identifying methodology"[59] (CIM) as an approach to exploring some fundamental questions regarding the nature of human landscape preference:

> What do people find salient in a given scene? What is it that results in a whole group of scenes being responded to in a similar fashion? There are many ways to categorize a particular environment. Use of content-identifying methodologies and preference ratings by untrained participants yields categorizations that are distinctly different from those generated by various professionals. The meaningful groupings identified permit comparisons across diverse studies.[60]

The CIM employs nonmetric factor analysis and hierarchical cluster analysis as analytical tools to explore the content of scenes that have been rated for visual preference by a group of respondents. Clearly this represents quite a different approach to the problem of landscape preference than the psychophysical approaches discussed earlier. The CIM stresses the identification of categories of environmental content such as spatial enclosure, depth of the scene, or blocked views. When the same method is employed for other landscape contexts, and the same categories emerge, it is possible to generalize across scenes. In other words, spatial enclosure may be the result of pine trees or deciduous trees; it is not the type of tree which is important here as much as their role in enclosing the scene. In the psychophysical approach, it is possible that no similarity between deciduous and coniferous forests could be discovered as the coefficients for each may differ each time measurements are made as was the experience of Brown and Daniel[61] in their studies of ponderosa pine forests.

Kaplan upon reviewing a number of examples of studies using CIM concludes:

> ... there appears to be both an empirical and a theoretical basis for categorizing landscape scenes. As is often the case with a satisfying research experience, these categories would have been hard to anticipate, but in retrospect make intuitive sense. These findings may also play a useful role in the further development of the landscape assessment research. In a recent review Stokols (1978) argues that generalizing findings across different settings requires a "theoretically based taxonomy of environments." The identification of consistent and interpretable patterns across a variety of different settings constitutes a first step toward developing such a taxonomy.[62]

The Kaplans, based on years of research using CIM, have developed and refined an information processing theory of landscape preference.[63] The theory is based on the concept that landscape preference must have evolved as an adaptive process as humans developed the mental and perceptual capacities for processing visual information important for survival. From this perspective, environmental preference is made up of two basic human needs for information: making sense and involvement. These two informational needs are also set in a time (present and future) and spatial (2-dimensional and 3-dimensional) framework which comprise four factors in the landscape preference model as shown in Table 3,10.

This model has been explored in a number of studies in the U.S. and abroad. Ulrich[65] in a study of Americans and Swedish groups found the information processing model was a valid basis for predicting the visual preference of both groups. In particular, Ulrich found complexity, focality, ground surface texture, depth and mystery as component variables for measuring aspects of the Kaplans' information processing model.

Brown and Itami[66] developed a visual resource management procedure as part of a regional planning study in Australia. In that procedure, the Kaplans' information processing model was used as a rationale for interpreting the findings of psychophysical research[67] to select variables of landform and landcover in a procedure to assess scenic quality. Research

Table 3,10 INFORMATION PROCESSING MODEL OF LANDSCAPE PREFERENCE[64]

Time/Spatial Context	Informational Needs	
	Making Sense	**Involvement**
Present/2-Dimensional	Coherence	Complexity
Future/3-Dimensional	Legibility	Mystery

by Herbert[68] tested the validity of the Brown/Itami model on a similar landscape in Michigan and found support for the use of the information processing model as a framework for scenic quality models. Gimblett *et al.*[68] investigated the mystery component of the model in rural settings in Ontario, Canada and found it to be a consistently perceived attribute of landscapes and were able to identify specific, measurable landscape attributes which contribute to the perception of mystery. This work, which draws on the expert, psychophysical and cognitive paradigms as a basis for designing improved visual assessment models tries to draw on the strengths of each.

The integration of approaches has also been shown to be profitable in the assessment of visual preference for Louisiana River landscapes. In that study, Lee[70] developed a visual assessment model which derived numeric scores for scenic quality based on composited ratings for the four components of Kaplans' information processing model (along with spatial definition, distinctive elements and disturbance factors). The model was applied to a series of landscape scenes and the resulting scores were compared statistically with ratings of landscape preference for the same scenes. Lee found good correspondence between preference ratings and ratings from the visual assessment model, again lending support to the information processing model of landscape preference.

The value of the cognitive approach, as demonstrated by the Kaplans and their associates, is in its explicit theoretical framework which attains generality without sacrificing reliability and validity. By utilizing some of the regression models of the psychophysical approach, it should be possible to relate physical landscape variables to the cognitive dimensions (i.e. complexity, coherence, legibility, and mystery), thus improving the utility and generalizability of both approaches. This conclusion is supported by Porteous[71] and Daniel and Vining.[72] Research in both paradigms has led to a more concrete understanding of the nature of visual preference for landscapes. It appears that, by combining the cognitive and psychophysical approaches, technical standards can be achieved. How can these research findings be incorporated into applied visual assessment and visual impact procedures.

THE BABY AND THE BATH WATER

Scenic assessment and the study of landscape perception has evolved rapidly over the last twenty years. Recent reviews of the literature have shown it is a growing field.[73] There is little evidence to believe that either all the questions or all the answers regarding landscape perception have been addressed. A case in point is recent work into the perception of different user groups. Zube *et al.*[74] point out that most landscape research has drawn its subjects from a restricted age span with respondents drawn largely from young and middle-aged adults. They point to research showing important cross cultural differences in populations. In their own research, they found distinct differences between children and older adults to young and middle-aged adults in perceptions of scenic quality and the landscape factors which contribute to scenic quality. Clearly future landscape research and applied landscape assessment techniques must address the differences among age and cultural groups.

As the field matures, there is good reason to continually improve applied visual assessment techniques. Because of the growing perspective on the field of landscape perception, caution should be taken in imposing rigid standards. Though they favour the implementation of standards for landscape assessment procedures, Craik and Feimer warn:

> We cannot allow technical standards set at premature or unrealistic levels to generate discouragement about the entire effort or to offer an excuse for dismissing these considerations out of hand. Instead, we require a perspective which shows that the specification and application of technical standards is a necessary and inevitable step in the maturation of effective visual resource and impact assessment procedures.[75]

In other words rigid scientific standards and overly complex techniques may disillusion practitioners and administrators who may then be tempted to throw out the baby with the bath water.

At present, the achievement of reliable, valid assessments of visual preference across a broad spectrum of landscape types and observers requires the use of psychological testing procedures on a case by case basis. The reasons for this are twofold. First, landscape perception research is not conclusive for a broad range of landscapes and observer groups. Second, existing applied visual assessment and impact procedures do not, according to recent studies[76], meet standards of reliability and validity.

The prospect of employing psychological testing techniques as part of applied visual assessment procedures is attractive from an academic view since this approach would rapidly increase the body of knowledge concerning landscape perception across a broad range of landscapes and people. However, from a practical viewpoint, Smardon *et al.* suggest that in many cases this level of rigour is not warranted. They state:

> The sophistication of VIA should be comparable to the complexity, importance or controversy of the project in question. For most projects a simple one-page rating form should suffice, especially if the project is typical and is structural in nature.[77]

There are also some practical limitations in employing rigourous psychological tests in applied settings because of the level of expertise in field personnel. Laughlin[78] found that only 20 percent of the VMS practitioners in the USDA Forest Service had education above the bachelor's degree level and there are no Ph.D.'s in that group. This suggests that there is little research expertise in this group. This need not be a hinderance with proper training in the use of psychological testing procedures. Consulting psychologists could then monitor the tests and assist in the interpretation of the results.[79]

Dearden[80] and Kaplan and Kaplan[81] argue for the integration of visual assessment techniques with public participation. Both Dearden and the Kaplans acknowledge that this approach is problematic, but argue these problems can be worked out much as research approaches in general must be tested and refined. In essence, public participation is viewed as applied landscape preference research. Brown and Daniel[82] argue that, because psychological methods produce quantitative measures of public preference, it improves the integration of aesthetic factors with other resources in the management process. Since the methods are reliable and valid they also improve the justification of decisions and the relationship between the client (the public) and the landscape architect.

Does this approach suggest the destruction of existing visual management systems? Probably not. The structure of the existing visual management procedures address issues which are important in the context of multiple use decision making. These procedures have been static however and have benefited little from the growing body of landscape perception research. By incorporating psychological testing procedures as part of a public participation program in the visual management system, research results can have direct influence on applied visual management systems and the body of knowledge on public landscape preferences would accelerate. In addition, if public preferences change in the socio-cultural evolutionary sense described by Craik and Feimer,[83] the public participation framework would provide a way of monitoring these changes.

SOME SPECIFIC SUGGESTIONS

Most of the landscape perception research reviewed here has concentrated on aspects of scenic quality or visual impact assessments. The research by Laughlin[84] and Grden[85] show clear concern by visual managers for the public sensitivity aspect of the visual management procedure. On comparing applied visual assessment and impact procedures with findings from psychological research, some specific observations and suggestions can be made regarding the integration of public perception research into a public participation approach to visual resource management.

First, the landscape classification process can be used to identify key changes in landscape character to assure complete photographic documentation of the landscape under study.[86] These photos not only provide valuable records of landscapes for the purpose of monitoring landscape change but can also be used in landscape perception tests as visual stimuli.

Second, user groups should be carefully identified and encouraged to participate in the visual management process. This not only serves to meet the requirements of visual impact assessment but provides an important cross-section of participants for preference testing. These user groups should represent a good range of interest, age and cultural groups.

Third, using well-tested methods of obtaining landscape preference scores, reliable, valid, quantitative scenic quality ratings may be obtained. This data can then be analyzed using the content identifying methodology to determine significant patterns in perception and to identify salient physical variables for extrapolating scenic quality ratings to other landscape in the study site. This will provide the data necessary to produce a reliable map of scenic quality.

Fourth, in addition to scenic quality ratings, valuable information relating to the public sensitivity procedures may also be obtained using questionnaires. Information such as the cultural significance of certain landscape features, the degree of concern for landscape changes, intensity of use, and types of use can provide valuable information with the same levels of reliability and validity as with the scenic quality ratings.

Fifth, with contentious issues, good photographic simulations of landuse proposals can be used to get quantitative measures of impacts on scenic quality using the same techniques to obtain the initial scenic quality ratings.

When this information is systematically incorporated with distance zones and viewshed analyses, definitive statements can be made regarding the existing character and quality of the landscape, the magnitude of visual impact of landuse proposals, and the probable public reaction to these projects. The process need not be onerous either to the visual management specialists or to the public participants. A well designed research/participation program can be enlightening and fruitful to both groups (see suggestions by Kaplan and Kaplan[87]). By including a free-response component in the process, fresh insights into the study of landscape perception may be discovered. This also provides opportunities for incorporating the orientation of the experiential paradigm into the visual management process.

CONCLUSION

This paper has reviewed recent landscape perception research which has bearing on applied visual resource management systems. Though there has been a growing volume of public preference research, this work has had little influence on applied visual assessment and visual impact studies. This can be attributed to the rather high satisfaction ratings practitioners give existing visual management systems[88] and their lack of research training. In addition, psychological approaches to landscape perception have not been conclusive across a broad range of landscapes or observer groups.

The psychophysical paradigm and the cognitive paradigm have produced landscape preference research along different conceptual lines. The psychophysical approach relates physical landscape variables to human preferences for scenery, but these results are not generalizable to a broad range of landscape types. The cognitive approach has concentrated on generalizable theories of landscape preference and has shown promise as a valid framework in applied visual assessment procedures. These two approaches are not alternatives, but can be complementary. This paper argues that an effective, and profitable way of incorporating landscape perception research into applied visual management systems is through a public participation process that works within the framework of existing visual resource management systems.

The USDA Forest Service and the USDI Bureau of Land Management are currently in the process of reviewing their respective visual resource management systems.[89] The U.S. Army Corps of Engineers are reviewing a draft proposal of a new visual impact procedure from State University of New York.[90] These actions show a continued interest by the U.S. Federal government in the refinement of institutionalized visual management procedures. The challenge before these agencies will be to develop strategies for improving the reliability and validity of their visual management procedures while maintaining their current utility. The solutions may not be simple, but there now exists techniques and expertise for achieving technical standards in visual assessment. The implementation of landscape perception research techniques into a public participation program in existing visual resource management systems can provide long term benefits for both the resource decision makers and study of landscape perception.

ACKNOWLEDGEMENTS

The author would like to gratefully acknowledge the helpful discussions with: Robert Ross, USDA Forest Service, Washington D.C.; Larry Isaacson, Department of Transport, Federal Highway Administration, Washington D.C.; Terry Daniel, Psychology, University of Arizona; and Richard Smardon, State University of New York, Syracuse in providing background information for this paper. Also Ervin Zube, University of Arizona provided helpful comments on an earlier draft of this paper. The ideas expressed in this paper are solely the responsibility of the author and do not necessarily reflect the opinions of the people or agencies mentioned.

REFERENCES

1. LEWIS, P.H. Jr., "Quality Corridors for Wisconsin", *Landscape Architecture*, Vol. 54, 2, 1968, pp. 100-107; LITTON, R.B. Jr., *Forest Landscape Description and Inventories — A Basis for Land Planning and Design*. Berkeley: Research Paper, PSW-49, Pacific Southwest Forest and Range Experiment Station, U.S. Department of Agriculture Forest Service, 1968; SARGENT, F.O., *Scenery Classification*, Vermont Resources Research Center, Vermont Agricultural Experiment Station, Report 18, 1967.

2. SMARDON, R.C., "When is the Pig in the Parlor?: The Interface of Legal and Aesthetic Considerations", *Environmental Review*, Vol. 8, 2, 1984, pp. 147-161; ZUBE, E.H., "Scenery as a Natural Resource: Implications of Public Policy and Problems of Definition, Description and Evaluation", *Landscape Architecture*, Vol. 63, 2, 1973, pp. 126-132; ZUBE, E.H., *Environmental Evaluation: Perception and Public Policy*. Cambridge University Press, University of Cambridge, 1980.

3. ZUBE, E.H., SELL, J.L. and TAYLOR, J.G., "Landscape Perception: Research, Application and Theory", *Landscape Planning*, Vol. 9, 1, 1982, pp. 1-33.

4. DANIEL, T.C. and VINING, J., "Methodological Issues in the Assessment of Landscape Quality", in Altman, I. and Wohlwill, J.F. (eds.), *Human Behaviour and Environment, Vol. 6*. New York: Plenum Press, 1983, pp. 39-84.

5. ZUBE, E.H. *et al*, *op. cit.*, 1982, p. 8.

6. DANIEL, T.C. and VINING, J., *op. cit.*, p. 56.

7. CRAIK, K.H. and FEIMER, N.R., "Setting Technical Standards for Visual Assessment Procedures", in Elsner, G.H. and Smardon, R.C. (technical co-ordinators), *Proceedings of Our National Landscape: A Conference on Applied Techniques for Analysis and Management of the Visual Resource*, USDA Pacific Southwest Forest and Range Experiment Station, General Technical Report PSW-35, Berkeley, California, 1979, pp. 93-100.

8. ZUBE, E.H. *et al, op. cit.*, 1982, p.8.

9. KAPLAN, S. and KAPLAN, R., *Cognition and Environment: Functioning in an Uncertain World*. New York: Aeger, 1982.

10. APPLETON, J., *The Experience of Landscape*. New York: John Wiley and Sons, 1975.

11. ZUBE, E.H. *et al, op. cit.*, 1982.

12. DANIEL, T.C. and VINING, J., *op. cit.*

13. *Ibid.*, p. 72.

14. USDA FOREST SERVICE, "The Visual Management System", *National Forest Landscape Management*, Vol. 2, Ch. 1, Agriculture Handbook Number 462, Washington, D.C., 1974.

15. USDI BUREAU OF LAND MANAGEMENT, *Visual Resource Management Program*, BLM, Division of Recreation and Cultural Resources, Washington, D.C., 1980; USDI BUREAU OF LAND MANAGEMENT, *Visual Simulation Techniques*, BLM Division of Recreation and Cultural Resources, Washington, D.C., 1980; USDI BUREAU OF LAND MANAGEMENT, "8400 — Visual Resource Management" (revision, release 8-24), Washington, D.C., 1984.

16. DOT FEDERAL HIGHWAY ADMINISTRATION, *Visual Impact Assessment for Highway Projects*, Contract DOT-FH-11-9694, American Society of Landscape Architects, Washington, D.C., no date.

17. FENNEMEAN, N.M., "Landscape Evaluation: A Research Project in East Sussex", *Regional Studies*, Vol. 2, pp. 41-55.

18. USDA FOREST SERVICE, *op. cit.*, pp. 5-6.

19. DOT FEDERAL HIGHWAY ADMINISTRATION, *op. cit.*, pp. 31-33.

20. USDI BUREAU OF LAND MANAGEMENT, *Visual Resource Management Program, op. cit.*

21. USDI BUREAU OF LAND MANAGEMENT, "8400 — Visual Resource Management", *op. cit.*

22. USDA FOREST SERVICE, *op. cit.*, p. 12.

23. DOT FEDERAL HIGHWAY ADMINISTRATION, *op. cit.*, pp. 13, 28.

24. USDA FOREST SERVICE, *op. cit.*, p. 18.

25. USDA FOREST SERVICE, *op. cit.*

26. USDI BUREAU OF LAND MANAGEMENT, *Visual Resource Management Program*, *op. cit.*

27. ANDERSON, L., MOSIER, J. and CHANDLER, G., "Visual Absorption Capability" in Elsner, G.H. and Smardon, R.C. (technical co-ordinators), *op. cit.*, pp. 164-171.

28. CRAIK, K.H. and FEIMER, N.R., *op. cit.*

29. LAUGHLIN, N.A., "Attitudes of Landscape Architects in the USDA Forest Service Toward the Visual Management System", unpublished MLA thesis, School of Renewable Natural Resources, University of Arizona, Tucson, Arizona, 1984.

30. *Ibid.*, p. 39.

31. *Ibid.*, p. 40.

32. *Ibid.*, P. 41.

33. *Ibid.*, p. 31.

34. GRDEN, B.G., "Evaluation and recommendations concerning the visual resource inventory and evaluation systems used within the Forest Service and the Bureau of Land Management", in Elsner, G.H. and Smardon, R.C. (technical co-ordinators), *op. cit.* pp. 296-304.

35. LAUGHLIN, N.A., *op. cit.*

36. CRAIK, K.H. and FEIMER, N.R., *op. cit.*, p. 95.

37. SMARDON, R.C., "Final Project Report: Development of Visual Activity Classification and Advance Testing on Visual Impact Assessment of Manual Procedures PSW Co-operative Agreement PSW-80-0005", submitted to Litton, R. Burton Jr., Land Use and Landscape Planning Methodology Research Unit, Pacific SW Forest and Range Experiment Station, Berkeley, California, 1981.

38. DANIEL, T.C. and VINING, J., *op. cit.*, p. 40.

39. KOPKA, S. and ROSS, M., "A Study of the Reliability of the Bureau of Land Management Visual Resource Assessment Scheme", *Landscape Planning*, Vol. 11, 1984, pp. 161-166.

40. MILLER, P.A., "A Comparative Study of the BLM Scenic Quality Rating Procedure and Landscape Preference Dimensions", *Landscape Journal*, Vol. 3, 2, 1984, pp. 123-135.

41. *Ibid.*, p. 123.

42. SMARDON, R.C., FEIMER, N.R., CRAIK, K.H., and SHEPPARD, S.R.J., "Assessing the Reliability, Validity and Generalizability of Observer-based Visual Impact Assessment Methods for the Western United States", in Rowe, R.D. and Chestnut, L.G. (eds.), *Managing Air Quality and Scenic Resources at National Parks and Wilderness Areas*. Boulder, Colorado: Westview Press, 1983, pp. 84-101.

43. CRAIK, K.H. and FEIMER, N.R., *op. cit.*; FEIMER, N.R., SMARDON, R.C. and CRAIK, K.H., "Evaluating the Effectiveness of Observer-based Visual Resource and Impact Assessment Methods", *Landscape Research*, Vol. 6, 1, 1981, pp. 12-16.

44. DANIEL, T.C. and BOSTER, R.S., *Measuring Landscape Aesthetics: The Scenic Beauty Estimation Method*, USDA Forest Service Research Paper Rm-167, Rocky Mountain Forest and Range Experiment Station, Fort Collins, Colorado, 1976.

45. SMARDON, R.C. *et al.*, *op. cit.*, pp. 94-95.

46. *Ibid.*, p. 98.

47. *Ibid.*

48. DANIEL, T.C. and BOSTER, R.S., *op. cit.*

49. DANIEL, T.C. and SCHROEDER, H., "Scenic Beauty Estimation Model: Predicting Perceived Beauty of Forest Landscapes" in Elsner, G.H. and Smarden, R.C. (technical co-ordinators), *op. cit.*, pp. 514-523; BUHYOFF, G.J., WELLMAN, J.D., and DANIEL, T.C., "Predicting Scenic Quality for Mountain Pine Beetle and Western Spruce Budworm Damaged Vistas" *Forest Science*, Vol. 28, 1982, pp. 827-838.

50. DANIEL, T.C. and BOSTER, R.S., *op. cit.*

51. BROWN, T.C. and DANIEL, T.C., *Modeling Forest Scenic Beauty: Concepts and Application to Ponderosa Pine*, USDA Forest Research Paper RM-256, Rocky Mountain Forest and Range Experiment Station, Fort Collins, Colorado, 1984.

52. IM, SEUNG-BIN, "Visual Preference in Enclosed Urban Spaces: An Exploration of a Scientific Approach to Environmental Design", *Environment and Behaviour*, Vol. 16, 2, 1984, pp. 235-262.

53. ANDERSON, L.M., and SCHROEDER, H.W., "Application of Wildland Scenic Assessment Methods to the Urban Landscape", *Landscape Planning*, Vol. 10, pp. 219-237.

54. DANIEL, T.C. and VINING, J., *op. cit.*, p. 79.

55. SMARDON, R.C. *et al.*, *op. cit.*

56. APPLETON, J., *op. cit.*; APPLETON, J., "Pleasure and the Perception of Habitat: A Conceptual Framework", in Sadler, B. and Carlson, A. (eds.), *Environmental Aesthetics: Essays in Interpretation*. Victoria: University of Victoria, Department of Geography, 1982, pp. 27-46.

57. KAPLAN, S. and KAPLAN, R., *op. cit.*

58. BROWN, T.J. and ITAMI, R.M., "Landscapes Principles Study: Procedures for Landscape Assessment and Management — Australia", *Landscape Journal*, Vol. 1, Fall, 1982, pp. 113-121; HAMMITT, J.E., "Assessing Visual Preference and Familiarity for a Bog Environment" in Smardon, R.C. (ed.), *The Future of Wetlands*, Towawa, New Jersey: Allanheld, Osmun Publishers,

1983, pp. 217-226; LEE, M.S., "Landscape Preference Assessment of Louisiana River Landscapes: A Methodological Study", in Elsner, G.H. and Smardon, R.C. (eds.), *op. cit.*, pp. 572-580; LEE, M.S., "Assessing Visual Preference for Louisiana River Landscapes", in Smardon, R.C. (ed.), *op. cit.*; ULRICH, R.S., "Visual Landscape Preference: A Model and Application", *Man-Environment Systems*, Vol. 7, 5, 1977, pp. 279-293.

59. KAPLAN, S., "Concerning the Power of Content-identifying Methodologies", *Assessing Amenity Resource Values*, USDA Forest Service General Technical Report RM-68, Rocky Mountain and Range Experiment Station, Fort Collins, Colorado, 1979a.

60. *Ibid.*, p. 13.

61. BROWN, T.C. and DANIEL, T.C., *op. cit.*

62. KAPLAN, S., *op. cit.*, p. 13.

63. KAPLAN, S. and KAPLAN, R., *op. cit.*; KAPLAN, S., "Perception and Landscape: Conceptions and Misconceptions", in Elsner, G.H. and Smardon, R.C. (eds.), *op. cit.*, pp. 241-248; KAPLAN, S. and KAPLAN, R., *op. cit.*

64. KAPLAN, S. and KAPLAN, R., *op. cit.*

65. ULRICH, R.S., *op. cit.*

66. BROWN T.C. and ITAMI, R.M., *op. cit.*

67. ZUBE, E.H., PITT, D.G. and ANDERSON, T.W., *Perception and Measurement of Scenic Resources in the Southern Connecticut River Valley*. Amherst, Mass.: Institute for Man and Environment, University of Massachusetts, 1974; ANDERSON, T.W., ZUBE, E.H., and MACCONNELL, W.P., "Predicting Scenic Resource Values", in Zube, E.H. (ed.), *Studies in Landscape Perception*. Amherst, Mass.: Institute for Man and Environment, University of Massachusetts, 1976, pp. 6-69.

68. HERBERT, J.E., "Visual Resource Analysis: Preference and Prediction in Oakland County, Michigan", unpublished MLA Thesis, University of Michigan, Ann Arbor, Michigan, 1981.

69. GIMBLETT, H.R., ITAMI, R.M., and FITZGIBBON, J.E., "Mystery in an Information Processing Model of Landscape Preference", *Landscape Journal*, Vol. 4, 2, 1985, pp. 87-95.

70. LEE, M.S., *op. cit.*, 1979; LEE, M.S., *op. cit.*, 1983.

71. PORTEOUS, D., "Approaches to Environmental Aesthetics", *Journal of Environmental Psychology*, Vol. 2, 1982, pp. 53-66.

72. DANIEL, T.C. and VINING, J., *op. cit.*

73. ZUBE, E.H. *et al*, *op. cit.*, 1982.

74. ZUBE, E.H., PITT, D.G., and EVANS, G.W., "A Lifespan of Developmental Study of Landscape Assessment", *Journal of Environmental Psychology*, Vol. 3, 1983, pp. 115-128.

75. CRAIK, K.H. and FEIMER, N.R., *op. cit.*, p. 96.

76. SMARDON, R.C. *et al*, *op. cit.*; KOPKA, S. and ROSS M., *op. cit.*; MILLER, P.A., *op. cit.*

77. SMARDON, R.C. *et al*, *op. cit.*, p. 100

78. LAUGHLIN, N.A., *op. cit.*

79. CRAIK, K.H. and FEIMER, N.R., *op. cit.*

80. DEARDEN, P., "Public Participation and Scenic Quality Analysis", *Landscape Planning*, Vol. 8, 1981, pp. 3-19.

81. KAPLAN, S. and KAPLAN, R., *op. cit.*

82. BROWN, T.C. and DANIEL, T.C., *op. cit.*

83. CRAIK, K.H. and FEIMER, N.R., *op. cit.*

84. LAUGHLIN, N.A., *op. cit.*

85. GRDEN, B.C., *op. cit.*

86. SMARDON, R.C. *et al*, *op. cit.*

87. For suggestions, see: KAPLAN, S. and KAPLAN, R., *op. cit.*, part 5.

88. GRDEN, B.G., *op. cit.*; LAUGHLIN, N.A., *op. cit.*

89. Personal conversations with: Robert Ross, USDA Forest Service, Washington, D.C.; Terry Daniel, University of Arizona; Del Price, USDI Bureau of Land Management, Washington, D.C.

90. STATE UNIVERSITY OF NEW YORK COLLEGE OF ENVIRON— MENTAL SCIENCES AND FORESTRY, "Visual Impact Assessment Procedure for U.S. Army Corps of Engineers, First Draft", Syracuse, New York, 1984.

PLATE 22 The Yorkshire Dales, England. ►

11 LANDSCAPE RESEARCH AND PRACTICE: RECENT DEVELOPMENTS IN THE UNITED KINGDOM

Edmund C. Penning-Rowsell

INTRODUCTION

Landscape research in the United Kingdom today appears both disparate and directionless, yet hardly can there have been a time when landscape considerations had so much popular concern or media coverage. This concern is reflected by events such as the major controversy over the publication of Marion Shoard's *The Theft of the Countryside*;[1] the bitter struggles between farmers and conservationists on issues such as drainage of wetland sites, notably the Halvergate Marshes;[2] the controversies over the erosion of green belt areas by developers encouraged by an anti-interventionist, anti-planning, Thatcher government; and continuing 'battles' over motorway routes through National Parks and other scenic areas. Add to those the recent year-long public inquiry over nuclear power plant siting and there appears to be any number of major issues where landscape protection or conservation is a central area of U.K. public debate.

The relative weakness of landscape research in Britain today, however, is partly illusory and partly real. The illusion occurs because many valuable studies escape the overall 'landscape' label by being hidden within the research efforts of single disciplinary areas. Other innovative and valuable research languishes on the shelves of planning departments and government agencies within unpublished internal reports. Nevertheless, the real state of landscape research — defined as having considerations of aesthetics at its core — is depressed precisely because the inter-disciplinary nature of this research makes progress more difficult given highly discipline-orientated research institutions. Moreover, the landscape research groups and formal associations in this field are still young and have yet to realize their initial aspirations let alone their full potential.

This chapter gives some insight into landscape research trends in Britain today, with an emphasis on the orientation of this research towards practical needs. It also includes an analysis of an agenda for future landscape research to compare with similar suggestions in this volume.

THE REALITY OF UK LANDSCAPE RESEARCH: TRENDS IN THE 1980s

The reality of landscape research, as opposed to aspirations and rhetoric, can be gauged in a number of ways, most of which are far from perfect. These include the published record found in theses, monographs and specialized academic journals, the funding of landscape research by government and other agencies, and the papers presented to conferences of practitioners dedicated to particular areas of landscape study. None of these is easily summarized to identify trends and progress. As identified above, this difficulty arises, at least in part, because so much landscape research is not conceptualized as such. It is undertaken by biologists or geographers, planners or ecologists within the framework, terms of reference and fashions of their discipline, using the language of research current there, and it often, therefore, lacks an explicit appeal to a wider disciplinary audience.

Central Government and Research Council Priorities

Government resources for research are either channelled indirectly to researchers through Research Councils — to their permanent Institutes or academics in higher education — or through direct contract from government departments to a variety of public agencies and private contractors.

The two U.K. Research Councils with an interest in landscape research are the Natural Environment Research Council (N.E.R.C) and the Economic and Social Research Council (E.S.R.C.). Perhaps surprisingly the funding from these Research Councils for landscape research, as defined here, is sparse. The Natural Environment Research Council has an overwhelming concern for the geological and biological sciences, and virtually no research is sponsored that is concerned with social aspects of landscape appreciation.[3] The Economic and Social Research Council, through its Environment and Planning Sub-Committee, has only supported a handful of small landscape-related topics in the last five years.[4] The most closely related topics tend to involve studies of land use change rather than landscape *per se*.

In terms of direct research funding by central government, much of the available resources — one way or another — is also focussed on landscape change. Priorities in recent years have placed particular emphasis on remote-sensed data collection and interpretation. The attempt to use modern satellite or other remote sensing technology and computer data processing in this way is designed to enable analysts to handle the massive amounts of information central to our understanding and appreciation of landscape character. Whether this ideal can be realized is open to question at this stage, but two studies serve to illustrate the trend in research technique and objectives.

Firstly, the Department of the Environment and the Countryside Commission are jointly sponsoring a major project for monitoring landscape change, undertaken by the aerial photography and remote sensing specialists Hunting Surveys and Consultants Limited. The objective of this study is:

> to obtain statistically reliable information on the current distribution, and extent, of features of major landscape importance in the countryside of England and Wales, and their rates of change.[5]

The incorporation of aesthetic value judgements is specifically eschewed. The data produced are intended to record landscape characteristics at a national level, because "regional and local data cut little ice when making a [... landscape conservation ...] case to central government".[6] The accuracy of the resulting data base should enable the detection of landscape changes of "around one percent per annum".[7]

In this way a national landscape data base will be prepared from available sources from which to monitor future landscape change. A combination of remote sensing and field survey (Figure 1,11) should gauge the accuracy of the former, and retrospective analysis of historic aerial survey data will give a longer-term context to the immediate results. Thus the project will provide "dependable information on historical change in the extent of such features (as) hedgerows, moorland and lowland heath over the last 30 years".[9]

As always with landscape monitoring, the crucial decision concerns the level of detail of the classification used to record landscape features, and the scale of these features to be recorded. Progress on this research project up until 1984 led to the development of a land use/landscape classification designed specifically for aerial photograph interpretation and associated field data collection. The classification was devised in preparation for Landsat Thematic Mapping covering sample areas in 31 out of the 54 counties in England and Wales.[10]

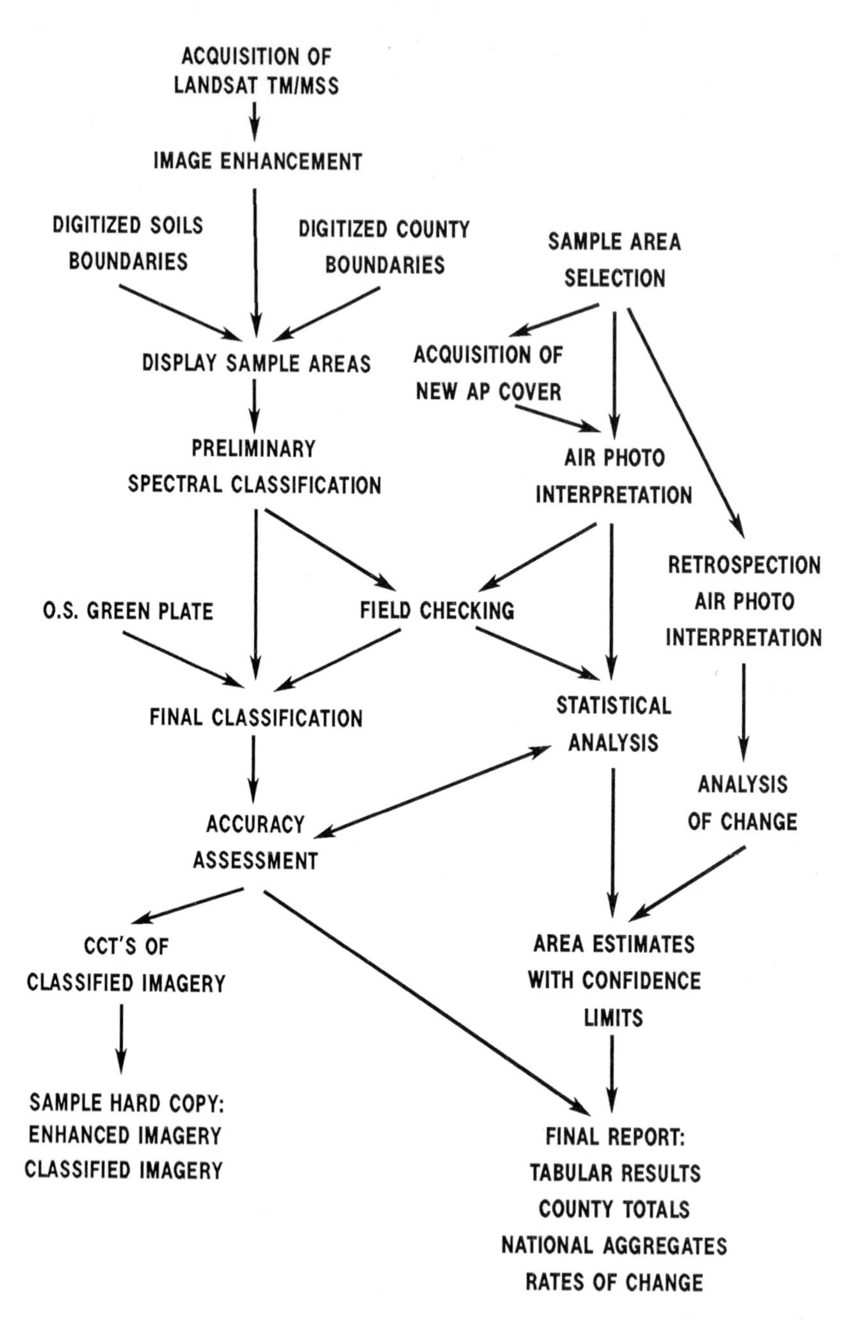

FIGURE 1,11 Stages in the monitoring of landscape change using satellite imagery.[8]

The land use/landscape classification that has been developed is surprisingly detailed (Table 1,11). Research has shown, however, that most of the units of the classification can be recognized during both aerial photographic and ground survey. However, some of the most detailed information, such as crop type and grass sward condition, can only be identified by fieldwork and scrub has proved to be the most difficult unit to interpret reliably using remote sensed data alone. Those limitations may not be of vital significance but could mean that the essentially qualitative differences in the 'health' of land uses — which are undoubtedly important to our aesthetic landscape judgements — are not measurable without time-consuming and expensive field surveys. Nevertheless, this project clearly promises much, even though the problems of monitoring landscape evolution are legion. An acute difficulty is that it may well be just those small-scale and highly detailed landscape features that cannot readily be recorded using modern technology which contribute most to landscape appreciation. Perhaps the key problem is that we simply do not know.

The second research project sponsored by central government focussing on remote sensing involves collaboration between the Welsh Office, the University of Nottingham and Wye College, University of London. The principal objective here is to determine whether Landsat II and other satellite imagery can be used to explore the complex landscape feature *combinations* that appear to be so essential to landscape understanding and appreciation. A secondary objective, dependent upon the first, is to map landscape classifications, based on these landscape feature combinations, to supersede a similar classification founded on a much more restricted data base also produced by the Welsh Office in 1980.[11]

Thus, for example, this project is exploring not just the ability of satellite imagery to recognize systematically certain landscape features, such as woodland and moorland, but also landscape characteristics such as the contiguous combination of woodland and moorland as recorded in adjacent image pixels. Field trials are being used to determine the accuracy of this recognition of what are essentially landscape 'assemblages'. Such is the power of modern computer processing that, when recognized accurately, these combinations of features can then be identified over large areas with little difficulty to provide a map of landscape character that is not just a record of land use alone. Furthermore, the combination of features can be mapped with imagery of different dates to provide a pattern of landscape evolution that is more subtle than an assessment of just the change in individual land cover types.

To take a second example, agriculture in much of upland Wales is economically unproductive and afforestation might be more worthwhile. To seek the most advantageous areas for commercial forestry, satellite imagery and other spatial data could be used to select from a large data base those

Table 1,11 LAND COVER CLASSIFICATION FOR AIR-PHOTO INTERPRETATION AND FIELDWORK FOR DEPARTMENT OF THE ENVIRONMENT/COUNTRYSIDE COMMISSION RESEARCH PROJECT MONITORING LANDSCAPE CHANGE IN ENGLAND AND WALES.[10]

	Code	Feature		Subclass
A.	**Linear Features:**			
	A1	Hedgerows		
	A2	Fences and insubstantial field boundaries		
	A3	Walls		
	A4	Banks - with or without low hedges		
	A5	Open ditches		
	A6	Woodland fringe		
	A7	Urban fringe edges		
B.	**Small or Isolated Features:***			
	B1	Isolated trees in hedgerows		
	B2	Isolated trees outside hedgerows		
	B3	Group of trees, mainly broadleaved (<.25ha)		
	B4	Group of trees, mainly coniferous (<.25ha)		
	B5	Linear features (strips of woody vegetation < 20m width and > 25m length)		
	B6	Farmland ponds		
C.	**Woodland:**			
	C1	Broadleaved high forest		
	C2	Coniferous high forest		
	C3	Mixed high forest (intimate mixture)		
	C4	Scrub (defined during fieldwork only)		
D.	**Semi-Natural Vegetation:**			
	D1	Upland heath	(a)	- ling (*Calluna*)
				- Bell heather (*Erica*)
			(b)	- Bilberry (*Vaccinium*)
	D2	Upland grass moor	(a)	- smooth grassland; fescues/bents (*Festuca, Agrostis*)
			(b)	- coarse grassland; purple moor grass (*Molinia caerulea*)
			(c)	- mat grass (*Nardus stricta*)
			(d)	- blanket bog; cotton grasses (*Eriophorum spp.*), peat-forming bog mosses (*Sphagnum spp.*)
	D3	Bracken		
	D4	Lowland heath	(a)	- rough grassland
			(b)	- heather
	D5	Gorse		
E.	**Farmed Land:**			
	E1		(a)	- ploughed/cropped land; cereals, ley grasses, legumes (crops identified during fieldwork only)
		Market Gardens	(b)	- including glass houses, nurseries and soft fruit farms
			(c)	- orchards
			(d)	- hops
	E2	Grassland	(a)	- improved pasture
			(b)	- rough pasture
			(c)	- neglected pasture

Table 1,11: (continued)

F.	**Water and Wet Lands:**		
	F1 Open Water		- coastal or estuarine
	F2 Open Water		- inland (not rivers)
	F3 Wetland vegetation	(a)	- peat bog (valley, raised, moss)
		(b)	- freshwater marsh (reed swamp)
		(c)	- saltmarsh
G.	**Other Land:**		
	G1 Non-vegetated peat		
	G2 Bare rock		
	G3 Sand		- dunes, dune-slack, shingle
	G4 Developed land	(a)	- built-up land; housing (including gardens), industrial, agricultural
		(b)	- urban open space; sports fields, parks, cemeteries
		(c)	- transport routes
		(d)	- quarries, mineral workings
		(e)	- derelict land; abandoned industrial sites and mineral workings.

* These categories to be analysed in relation to data from Forestry Commission sample strips only.

areas which are identical to currently afforested areas in all landscape features except for the woodland cover (i.e. altitude, slope, soil type, etc.). These areas are presumably those most likely to be favourable for tree planting but they can only be identified systematically by searching for landscape feature combinations, not just the features in isolation.

The essential prerequisite for this type of investigation is that the combinations of features that can be recognized from remote sensed data make sense 'on the ground'. This is where field investigations again are necessary to determine the combination of landscape features that create the essential character of distinctive landscapes. The combination of steep slopes, bare rock mixed with coniferous woodland gives a distinctive upland landscape whereas the same features without the rough topography or the woodland gives a landscape character which is altogether different. Thus land *use* recognition — as in the Department of the Environment/Countryside Commission project — is but the first stage towards land*scape* recognition, which is the objective of the Welsh Office/Nottingham University study.

The Published Record

To gauge further the topics of real and active concern to landscape researchers in Britain today, and gain some insight into research results rather than just aims and objectives, we can turn to what is being published in the significant research journals. This is more revealing than looking at the content of published books because, with some exceptions, these tend not to contain the most up-date-date research results.

The two journals selected here to represent British landscape research effort are *Landscape Research*, produced by the Landscape Research Group, and *Landscape Planning*, now known as *Landscape and Urban Planning*. The latter, of course, is Dutch but it is edited in England and has a wide British following. All the major papers in these journals were reviewed for the years 1979 to 1984 to identify both the character of landscape research as published and any trends through time. These aims are considerably assisted by the continuity of editorial control over both journals throughout the period.

The results are shown in Table 2,11, which has a number of interesting features. Firstly, the research reported in the two journals is not dissimilar in overall pattern. The major differences are that *Landscape Research* is more concerned with artistic research and scholarship than is *Landscape Planning*, which is more involved in policy analysis. Secondly, the aggregate results show the primary concern in both journals is the evaluation of official policy for landscape/planning/recreation/management. Clearly a major research interest concerns the success or otherwise of these policies, usually with an implicit or explicit view to promoting policy change. However, there appears, from the record of the last two years, to be waning interest in this area; perhaps the researcher's lack of influence upon policy is leading to less emphasis here. Another area of apparent waning interest concerns the whole topic of landscape analysis, classification and mapping. This contrasts with the government's concern in the field as gauged by research funding related to remote sensing as discussed above.

Thirdly, a 'main-stream' and continuing interest shown by the published record is the analysis of landscape preferences, taste and appreciation, both at a theoretical and an empirical level. Such is the wealth of results here that perhaps those who decry our lack of knowledge of landscape understanding have simply neglected the available literature.

Finally, the published record shows some interesting gaps and weaknesses. The link between landscape research and its exemplification in design is patchy. Research on the creation and impacts of new landscapes is sparse; the dearth of theory is striking. The 'everyday' landscapes of people's gardens and immediate local surroundings receive far less attention from researchers than the 'grand' landscapes of national parks and historic monuments.

Table 2,11 A CLASSIFICATION OF PAPERS IN *LANDSCAPE RESEARCH (LR)* AND *LANDSCAPE PLANNING (LP)* FROM 1979 TO 1984. NUMBERS (e.g. *LR9*) INDICATE VOLUME NUMBERS.

	1979		1980		1981		1982		1983		1984		Totals	
	LR4	*LP6*	*LR5*	*LP7*	*LR6*	*LP8*	*LR7*	*LP9*	*LR8*	*LP10*	*LR9*	*LP11*	*LR*	*LP*
1. General policy (27), landscape reclamation (2), recreation promotion, etc. (11), management (6), landscape and agriculture (8)	4	10	5	13	—	3	6	4	3	6	—	—	18	36
2. Literature and landscape (4); art and landscape (9)	2	—	—	—	1	—	3	—	—	—	7*	—	13	—
3. Preferences/taste/aesthetics: theoretical/review (12); empirical (20)	3	—	3	1	3	4	4	3	2	3	3	3	18	14
4. Landscape description, analysis classification, etc: narrative (3); classification, evaluation (12); computing/mapping (5); other (8)	1	3	4	2	8*	3	—	—	—	—	2	5	15	13
5. Ecological and biological analysis	—	—	3	1	6*	5	1	2	—	6*	—	1	10	14
6. Historical landscape evolution, historic landscapes, historic landscape reclamation and management.	3	5	—	2	—	—	6*	2	—	1	2	1	11	11
7. Design exemplification and education: theory (1); landscape reclamation (5); urban revival, city parks etc. and settlement design (10); general/education (4)	—	—	1	1	2	2	—	3	8*	—	—	3	11	9
TOTALS													96	97

* conference proceedings

Historic Landscapes: Practice With Research

A most noteworthy trend in both the published record and, more importantly, amongst some practising landscape architects and consultants is the growing volume of work on historic parkland and garden landscapes. This trend is perhaps part of the enduring British nostalgic concern for the past as a basis for landscape conservation policies today.[12] What is notable, however, is that at least some resources are now becoming available to pursue this area systematically — twenty years after the formation of the Garden History Society — which has lagged significantly behind concern for other government-designated parks and landscapes.[13]

This trend began in the 1970s and has been spearheaded by the National Trust and other enlightened landowners. A major landmark has been the compilation by the Historic Buildings Council of a data base in the form of the *Register of Gardens and Land of Special Historic Interest*;[14] unfortunately, as yet, this Register only covers England. Perhaps the most significant example of this type of landscape research and its link to practice is the analysis and 're-design' of the Capability Brown landscape of Blenheim Park, Oxfordshire.[15] Here, as elsewhere, the investigation and restoration plan has had to strike a balance between the different eras represented in the current park landscape (i.e. between the Wise and Vangurgh period, 1705-1760, the Capability Brown period, 1760-1790, and the era of the Ninth Duke of Marlborough, 1896-1935). Also, however, there are significant conflicts between current uses of the park which complicate its sensible restoration. A major park management objective is to grow trees for timber, which requires regular felling, but sporting interests require only minor felling at specific periods of the year so as to create the minimum of disturbance to game birds. The situation is compounded by years of decline in the parkland landscape when the necessary active management was neglected for financial reasons.

The restoration plan now devised envisages sequential tree planting on a continuing basis within distinct landscape zones reflecting their history and current functions. Selected felling is required within this plan, even of trees not yet mature, to open up vistas that formed part of the original design of the overall historic masterpiece. All the work in prospect is thus influenced by the most detailed historical research into the intended landscapes of the original designers[16] but is also significantly influenced by the uses of the park today and thus the current commercial and financial constraints. Figure 2,11 shows how the scholastic historical investigations and the detailed analysis of the current economics of the existing landscape are the essential complementary inputs to the practice of historic landscape restoration.

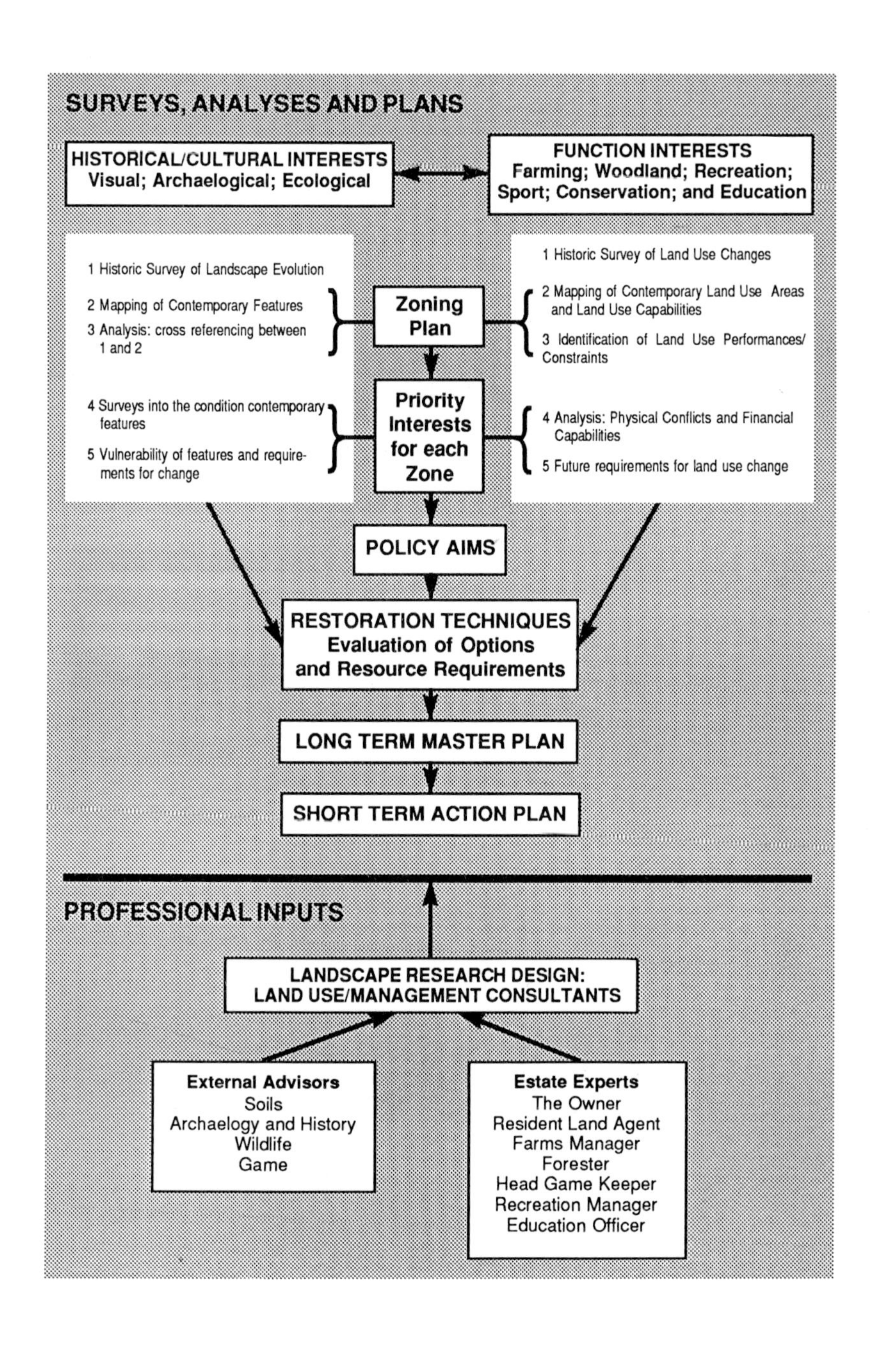

FIGURE 2,11 The processes involved in producing the landscape restoration plan for Blenheim Park, Oxfordshire, England.[17]

LESSONS FROM AN AGENDA FOR FUTURE LANDSCAPE RESEARCH

Against the background of the developments in landscape research literature and practice discussed above it is revealing to examine the topic areas priorized for future research effort by a knowledgeable group of landscape researchers and practitioners in England (see Appendix 1,11). The context for this identification of research needs was a conference held by the Landscape Research Group in London during April 1984 centred around the complexities of *Landscape Meanings and Values*.[18] The Landscape Research Group is unique in bringing together a wide range of specialists, including landscape architects and consultants, local and central government officials, pressure group activists, and academics in a number of disciplines including landscape architecture, geography, english and fine art.

Within this conference, a Working Party of 34 participants representing this diversity of interests sought to identify key issues in relation to future research needs. The sphere of inquiry was restricted to the fundamental area of the meaning of landscapes to people, and the bases for their valuation of specific landscapes and their features. This analysis of research needs led to the identification of major lacunae in our understanding of landscape, including landscape preferences, tastes, understanding, perception and utilization. Indeed, faced with the task of setting down the collective wisdom from several decades of experience in landscape research it was nevertheless sometimes difficult for the Working Party to define an important field in which much is known at all!

The results of this deliberation illuminate considerably the areas of investigation that are both needed by the practitioners — who numbered nearly forty percent of the Working Party — and those which appear to the academics/researchers as at least having some chance of implementation. To an extent, they also reveal the serious problems of communicating research results from researcher to practitioner. The latter is often ignorant of research material already available, presumably because it has been produced in a form not amenable to, or accessible for, use. Conversely, the academics are ignorant of the research that could usefully contribute to better practice.

Landscape History, Language and Art

What lessons can be learnt, then, from this complementary identification by researchers and practitioners of research needs and possibilities? Firstly, there is dissatisfaction with our theoretical understanding of landscape values, hence the call for more research into fundamental landscape concepts and meanings. This proposal transcends particular landscapes to embrace such themes as the concern for the symbols created in historic landscapes, the evolution of the landscaping tradition, and techniques for reconciling modern land uses with historic or aesthetic qualities.

As researchers pursue these more fundamental topics, especially within cross-disciplinary groups, we may realize the impoverished and ambiguous nature of the language of landscape. The very term 'landscape' is used differently across different cultures and within different professions. A landscape lexicon would enhance cross-disciplinary debate although it also might stifle initiative by over-standardizing the way we articulate landscape descriptions.

The role of the artist in creating landscape tastes needs exploring, as does the relation between the history of landscape art and landscape gardening/ architecture. In this way, we might deepen people's experience of current landscapes through art, which is inherently more accessible than the 'outdoor' landscape.

Landscape Taste: Preference and Prejudices

A second important point arising from Appendix 1,11 is that much more attention is needed to the fundamental topic of landscape taste, that is, preferences and prejudices for different landscape types.

This reflects a widespread concern that systematic studies of preferences are rare and tend to be restricted either to case studies which attempt to explain simplistically the basis of people's evaluation of particular landscape qualities[19] or to more literary expositions which the practitioner finds rather harder to apply.[20] It also reflects the growing theorization about landscape — attempting to look for 'biological' explanations of landscape preferences[21] — and the view that these hypotheses need empirical testing (if that is possible) so as to develop more generally applicable landscape theories. In addition, there is a concern that more is known about the landscape views and feelings of the literate and the more educated in society but little about the landscape values of the 'ordinary man' — whoever he or she may be — and little of the extent to which there is a consensus of landscape preferences in the population. Also, little is known about how landscape tastes are formed, both as we grow up and also in relation to landscapes we discover. How do we value new landscapes in relation to those more familiar and thus more commonly loved? The nuances of landscape values to be explored here are endless and the research in prospect is neither straightforward nor likely to lead to unambiguous results.

Education, Training and Design

More students in Britain in the 1980s are now studying landscape-related environmental studies than ever before. Our knowledge is sparse, however, of the landscape perceptions of the educators and the educated, and of the best ways of encouraging greater knowledge and awareness of landscape aesthetics and awareness amongst our emerging professional landscape designers. The problem is that outside the narrow landscape-related professions, such as town planning and landscape architecture, there is a host of different disciplines to be addressed. Even the current extent of environmental education and its curricular trends is unknown in Britain. The time is ripe for more systematic inventories of courses, programmes and the values they are designed to instill.

CONCLUSIONS

The links between research and practice are always problematic. In general, landscape researchers look for answers to fundamental theoretical questions. This takes time and effort over a number of years of concentrated research. In comparison, practitioners tend to need answers more quickly, and answers which can be applied directly to their immediate work.

What is clear from this assessment of research and practice in Britain is that the volume of research is small and practitioners are operating in a theoretical vacuum. The government promotes high technology-related landscape research and ignores the fundamental social nature of landscapes; few research resources are directed to social investigations of landscape meaning and values.

This is not to say that there is an absence of both need and ideas for future landscape research. The number of researchable topics is large and organizations such as the Landscape Research Group are now well positioned to co-ordinate a major programme to increase our landscape understanding and appreciation. This in turn may help practitioners to safeguard those landscapes about which public concern is greatest and to design new landscapes for future generations to enjoy.

Appendix 1,11 IMPORTANT RESEARCH TOPICS IDENTIFIED BY THE LANDSCAPE RESEARCH GROUP WORKING PARTY IN THE FIELD OF *MEANINGS AND VALUES IN LANDSCAPE.*

1. Collation of existing data, research material and bibliographies

- The production of a dictionary of landscape terminology.
- Assembling and sorting bibliographic material on landscape research.
- A biennial register of landscape research interests.
- Systematic identification of historic landscapes, their eligibility for conservation, and techniques for their management.
- A systematic search for 'archetypal' landscapes (e.g. the Devon landscape; the Fenland landscape).
- Assembling specific and detailed data on landscape painting and poetry.

2. Topics for research: historical/historic landscapes

- An investigation of the tradition of creating new landscapes.
- The relation between the biological and the historical in landscape meaning.
- The meaning and value of historic landscapes, including historic parks and gardens, and how they can be enhanced.
- Ways of reconciling modern land uses with historic/aesthetic qualities.
- A study of footpaths as cultural symbols: their use, value and relevance as heritage features.

3. Topics for research: theoretical, and/or the language of landscape

- The distinction between the concepts 'landscape' and 'environment'.
- Comparative cultural and historical study of the landscape idea and attitudes to nature.
- Collaborative research between the arts and sciences on symbolism in landscape.
- The relations between the economics and aesthetics of landscape gardening.
- Classification of division of landscapes by ridge lines.
- Exploration of the hierarchy of levels at which landscape is experienced (i.e. from innate biological to the political level).
- The relationship between landscape and political and economic forces.
- An analysis of the differences between taste and preferences and prejudices.
- An analysis of the language of landscape.
- The meaning of landscape: exploring its social and historical dimensions and implications for present day thought and practice.

4. Topics for research: arts and artistic studies

- How the different arts elicit responses to landscape. How is the sensual experience of one person 'reduced' to the verbal/audio/visual medium to create a sensual response in someone else.
- The extent to which landscapes compare qualitatively with the books, music, paintings and films we choose to remember them by.
- How are tastes transmitted from poets/writers/artists to 'ordinary people'?

5. Topics for research: apppreciation/experience studies of peoples' landscape attitudes and feelings

5.1 GENERAL

- A long-term census of changing landscape tastes (e.g. every five years).
- An analysis of the pre-conditioning influences on perception and appreciation of landscape.
- Human biological reactions to landscape.
- Personality assessments of key individuals involved in the landscape profession ('what makes us tick?').
- An exploration of personal versus communal taste.
- Studies of how people become acquainted with new places.
- Exploring the relationship between familiarity with landscapes and landscape values/meanings/attitudes.
- Analysis of the ways that contact with landscape deepens or distorts sensibility.
- The process by which landscape values are formed or transmitted (for a specific locality, e.g. Dartmoor).
- Values and meanings of 'everyday' landscapes to local residents.
- How people respond to the formal interpretation of landscapes.
- Research on the appreciation of designed (as opposed to 'natural' or 'wilderness') landscapes.
- Testing public landscape preferences by means of photographs, questionnaires, or market prices, to establish what features and assemblages are most highly valued and by whom.
- Analysing simulated preferred landscapes (see the work of the British Transport and Road Research Laboratory).

5.2 RELATED TO DIFFERENT POPULATION GROUPS

- How landscape meanings and values differ with differing experiences.
- How young children see and evaluate landscape.
- An analysis of landscape experience, and needs of different age groups.
- The class biasses of interest in landscape ('Is landscape middle class?').
- Social differences in taste, how much tastes diffuse from one group to another and how long such diffusion takes.

- How certain specific professional groups experience and value landscape (e.g. land agents).
- Evaluating methods of influencing the values of the land-based professions (i.e. farmers and land agents).
- *Who* likes *what* in terms of national and class differences.
- How far attitudes to landscape conservation differ from country to country (e.g. U.S.A. and U.K.).

5.3 RELATED TO SPECIFIC LANDSCAPES AND LANDSCAPE FEATURES

- What people expect from highly valued landscapes.
- The effects of transient phenomena on landscape appreciation.
- The meaning and value of water in the landscape.
- The role of diversity of form and character in our appreciation of landscapes: is aesthetic diversity derived from ecological diversity?
- The influence of debilitated 'problem' landscapes on the lives of people who live and work in them.
- An investigation of the appropriate meanings and values to be sought and obtained from the redeveloping inner city landscapes.
- Related to Appleton's (1975) prospect-refuge thesis, how do people respond when they cannot find prospect and refuge in a landscape?
- The way people re-value landscapes coming under threat.

6. Topics for research: design/management/interpretation/education and training

- Investigating methods and techniques for creating meaning and value in designed landscapes through the design process.
- Participation by researchers within an active and free-ranging design of landscape modifications (with an international dimension).
- An analysis of the extent to which presentations to decision-makers (e.g. to a Planning Committee) influence their decisions.
- The role of interpretation as a medium between the landscape and the contemporary observer.
- The relationship between environmental education and landscape perception.
- Evaluating the different ways of training students in landscape assessment.
- An evaluation of the best methods of including an awareness of landscape as a central part of training courses in architecture, landscape architecture, planning, farming, forestry, and estate management, etc.

REFERENCES

1. SHOARD, M., *The Theft of the Countryside*. London: Temple Smith, 1980.

2. PENNING-ROWSELL, E.C. *et al.*, *Floods and Drainage: British Policies for Hazard Reduction, Agricultural Improvement and Wetland Conservation*. London: George Allen and Unwin, 1986; O'RIORDAN, T., *Lessons from the Yare Barrier Controversy*. Norwich: University of East Anglia, School of Environmental Sciences, 1980.

3. NATURAL ENVIRONMENT RESEARCH COUNCIL (N.E.R.C), *Annual Report 1983/4*, Swindon, Wiltshire, 1985; INSTITUTE FOR TERRESTRIAL ECOLOGY (I.T.E.), *Annual Report 1983/4*, 1985.

4. ECONOMIC AND SOCIAL RESEARCH COUNCIL (E.S.R.C.), *Annual Report 1983/4*, London, 1985 (see also previous reports of the Social Science Research Council).

5. PHILLIPS, A., "Conservation at the crossroads: the countryside", *Geographical Journal*, 151, 2 (1985) p. 238.

6. *Ibid.*, p. 239.

7. *Ibid.*

8. *Ibid.*

9. *Ibid.*, p. 242.

10. HUNTING SURVEYS AND CONSULTANTS LIMITED, *Monitoring Landscape Change: Progress Report 2*. Borehamwood, Hertfordshire, 1984.

11. WELSH OFFICE, *A Landscape Classification of Wales*. Cardiff, 1980.

12. LOWENTHAL, D. AND BINNEY, M., *Our Past Before Us: Why Do We Save It?* London: Temple Smith, 1981; REED, M., (ed.) *Discovering Past Landscapes*, London: Croom Helm, 1985.

13. GOODCHILD, P., "The conservation of parks and gardens in the United Kingdom", *Landscape Research*, 9, 2 (1984) pp. 1-3.

14. HISTORIC BUILDINGS COUNCIL, *Register of Gardens and Land of Special Historic Interest*. London, 1984.

15. COBHAM, R., "Blenheim: the art and management of landscape restoration", *Landscape Research*, 9, 2, (1984) pp. 4-14; MOGGRIDGE, H., "The working method by which the original composition of planting around L. Brown's lakes was defined", *Landscape Research*, 9, 2, (1984) pp. 15-23.

16. MOGGRIDGE, H., *op. cit.*

17. *Ibid.*

18. PENNING-ROWSELL, E.C., AND LOWENTHAL, D. (eds), *Landscape Meanings and Values*. London: George Allen and Unwin, 1986.

19. For example see DEARDEN, P., "A statistical method for the assessment of visual landscape quality for land use planning purposes", *Journal of Environmental Management*, 9, (1980) pp. 51-68; PENNING-ROWSELL, E.C., "A public preference evaluation of landscape quality", *Regional Studies*, 16, 2 (1982), pp. 97-112.

20. For example see LOWENTHAL, D., *Finding Valued Landscapes*. Toronto: University of Toronto, Environmental Perception Working Paper 4, 1978.

21. For example see APPLETON, J., *The Experience of Landscape*, London: Wiley, 1975; ORIANS, G., "An ecological and evolutionary approach to landscape aesthetics", in PENNING-ROWSELL AND LOWENTHAL, *op. cit.*, p. 3-25.

Meares Island, just north of Pacific Rim National Park, British Columbia, was featured prominently in the advertising campaigns of the province (**PLATE 23**) and yet was slated for clearcutting before protesters stepped in (**PLATE 24**). ►

Summer in British Columbia offers more possibilities than you thought possible.

Say hello hello to echoing surf. Dive into bubbling hot springs and hit the heights in our high and mighty Rockies.

Then cruise through our fjordic coast that harbours Vancouver's super sights and natural wonders.

Our hotels are very accommodating and there's shopping galore.

Ahh, candlelight continental cuisine, a chopstick Chinatown feast or an impromptu picnic at your pick of over three hundred British Columbia parks.

Rent a houseboat and go putt putting down secluded silver lakes or peddle a bike up to a tasty roadside fruit stand and take a bite of the endless Okanagan sun.

Lure a Dolly Varden while fishing in extremely fruitful waters.

And remember, this year we're celebrating Funfest '80 with festive festivities.

What's more, British Columbia packages have vacation value all wrapped up so ask your travel agent for the inside story on the Inside Passage cruise, rugged outdoor adventures plus sensational city tours.

For fundamentals on Funfest '80 or our enchanting cities and great outdoors write Tourism British Columbia, Dept. 210, 1117 Wharf Street, Victoria, British Columbia V8W 2Z2.

Anything's possible in our Super, Natural Summer.

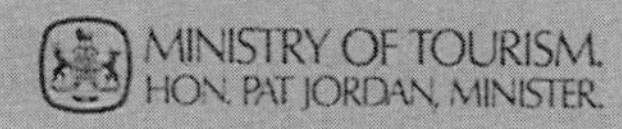

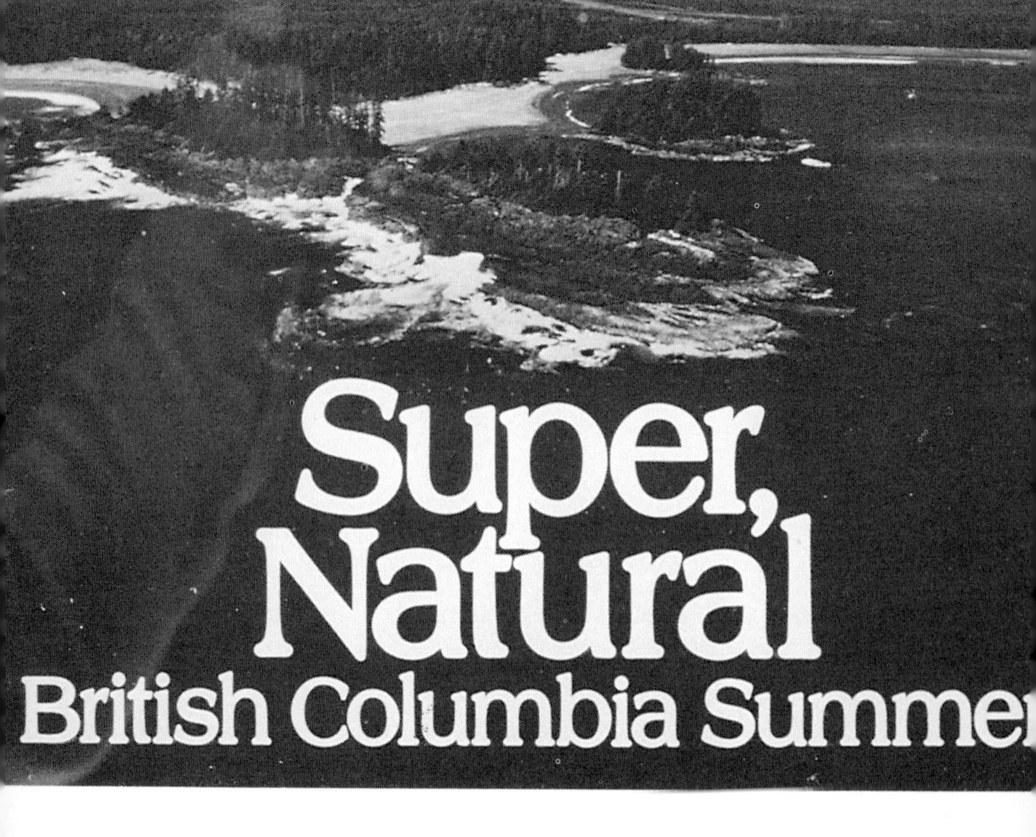

EDITORS' POSTSCRIPT

MEARES... IS NOT
"A CLEAR CUT ISSUE" YET.
UNITE and FIGHT

12 WHERE TO NOW? SOME FUTURE DIRECTIONS FOR GEOGRAPHICAL RESEARCH IN LANDSCAPE EVALUATION

Barry Sadler and Philip Dearden

Landscape evaluation, in the broadest sense of the term, is part of the geographer's training and vocation. The discipline, in the past, has been grounded in regional studies and comparisons of the material imprint of man-environment interaction over time and across cultures. Most geographers profess scholarly interest in this sweeping theme. The aesthetic dimension of landscape and place, whether evaluated directly or through the lens of everyday experience or artistic and literary response, is an intriguing part of geographical territory. It is, however, a subject for research specialization by only a small handful of practitioners, far outnumbered, for example, by those involved in the statistical manipulations of abstract data on spatial structure.

During the behavioural revolution of the 1970's, there was a burst of studies on landscape evaluation and related themes. This involved both the rediscovery and reworking of traditional concepts of regional, cultural and historical geography and the acquisition and adaption of a new set of tools and techniques through interdisciplinary contacts, notably with the psychological sciences. A diverse range of methodologies is now applied to the evaluation of landscape, ranging from the humanist and the phenomenological to the positivist and the experimental. Many of these are represented in this volume. The introduction has drawn attention to the opportunities for comparing the role and relative contribution of different approaches to landscape evaluation. In this postscript, the emphasis is on the next steps, on where we go from here in geographical research on landscape evaluation in Canada.

The preparation of this volume, from organization of the symposium to editing the papers involved contact with a number of geographers and others who are active in macroscale landscape evaluation in Canada and internationally. It helped crystallize certain perspectives on the course and tempo of research in this area. Our main conclusion, reflected in the literature, the convening of symposia and similar indicies of activity, is that landscape evaluation has stagnated in the past seven or so years in comparison to the previous decade. This assessment applies to Canada more than to the other countries to which comparisons are drawn in this volume. In the United States and the United Kingdom, our reading of the situation is that the level of interest in and output of landscape evaluation studies has been maintained at levels closer those which prevailed in the expansionary decade of research (c. 1970-1980).

Several reasons can be advanced to explain this difference. They include: 1) the tradition of safeguarding landscape amenity and character in British town and country planning; and 2) the more recent but relatively rapid incorporation of visual considerations and scenic concerns into the mandates of the U.S. Bureau of Land Management and the Forest Service which are collectively responsible for administering large tracts of the American West. By comparison, resource management institutions in Canada have neither inherited the British approach to amenity-based landscape planning nor adopted the systematic approach to visual analysis of U.S. public lands agencies. Landscape evaluation in Canada is still focussed at the site or microlevels rather than at macroscales. It is more the province of landscape architecture than of regional land use planning and resource management. The aesthetic and design currency of landscape architecture, moreover, also tends to be of minor importance in environmental impact assessment and related processes for development control. The work of public utilities in siting and routing large scale energy and transportation projects in rural and heritage lands is an exception. Even here, however, the role and place of aesthetics in environmental decision making is implicit rather than explicit; and it is hard, for example, to disentangle the relationship to ecological values, land use change and other resource management opportunity costs.

The marginal role and patchy application of landscape evaluation in resource management and development planning and control is both product and cause of the paucity of research and, by extension, the theoretical and methodological issues which underlie this field. A more concerted research effort needs to be mounted by geographers and others interested in the visual quality and aesthetic values of the Canadian landscape. Below, we have set out a preliminary research agenda for landscape evaluation. It is not intended to be comprehensive or definitve, rather it is meant to provide a frame of reference for, and an invitation to further discussion by geographers and others interested in this research area.

The agenda is organized into five sections. Under each category, there is a brief introductory statement followed by a short checklist of questions which, in our view, are worth further attention.

1. Philosophical and Theoretical

A great deal of attention has been devoted in the literature to examining the concept of aesthetics and its historical and spatial relationship to landscape, culture, and sensory and cognitive response. The philosophical understanding of what constitutes the aesthetics of landscape and its built and natural coordinates still varies widely among academic disciplines and the planning and design professions. We do not yet have a framework for evaluation, a system of reliable generalizations or formal principles of landscape aesthetics and closely related values and meanings. Such a framework should do more than statistically or qualitatively correlate spatial form and human response, whether expert or lay. It should incorporate and be founded on a theory of landscape aesthetics which explains why these certain types and components are valued, and why (rather than just how) they differ between different social groups and within the same culture over time.

How, in particular, might Canadian geographers draw from and apply the theoretical and philosophical insights being developed in and around the field of landscape evaluation to gain a better perspective on a fundamental question: What exactly is it we are trying to measure in studies of aesthetics, scenery, visual impact and so on? What are the archetypal qualities of the Canadian landscape, the elements which reflect and reinforce who we are and where we live? Are these amenable to categorization nationally, regionally, and thematically in terms of built, rural and wild landscapes? To what extent might these qualities organize or influence patterns of perception and response to everyday or exotic landscapes?

2. Historical and Contextual

Landscapes are mirrors of culture and society. They are made and remade by each successive generation of inhabitants. Spatial patterns of land use and occupance express the concrete and cumulative relationship of form and function. The morphology of landscape, at any point in time, is a dynamic balance of old and new elements, and of the forces of change and inertia. It reflects the ebb and flow of tides in taste and technology. The history of landscape cannot be fully evaluated just by tracing the evolution and diffusion of the form of settlement; it also involves understanding why people organized their living space as they did and how their values and images entered into and were shaped by this transaction. None of this is new to Canadian geographers, but they have yet to give such matters sustained attention and to provide aesthetic perspectives on the past landscapes as prologues of the present spatial order (or disorder according to certain critics).

What is the nature of the relationship between the spatial evolution of the Canadian landscape and paradigmatic shifts in images of nature, urbanism and rural life? How have "imported" values changed and been changed by Canadian landscapes? What role have the arts and letters played in structuring new axiologies of space? How has this influenced everyday responses to the Canadian landscape and *vice versa*? What, in the final analysis, are the Canadian traditions of landscape-making, including those of indigenous peoples as well as European settlers? And what is the contemporary relationship of the biological and historical in landscape meaning and symbolism?

3. Methodological and Topical

Methodological discussions of the relative utility of various tools and techniques for evaluating the aesthetic, scenic and sensory aspects of landscape are well developed. The arguments with respect to the role of lay and expert opinion, the importance of consensus, the value of quantification and so on have probably run their course given the present body of knowledge on landscape evaluation. A greater effort needs to be devoted to theory building and critical reinterpretation of research which spans the range from the phenomenological and subjective to the physical and objective. Within this context, there are important opportunities for building methodological orientations into ongoing substantive (policy or theory based) research. One approach might be an in-depth case study of a regional landscape which utilizes the range of tools to document its past and present character and associated values. This approach might also provide a pilot or demonstration project for application to other areas with further possibilities for comparative research.

What are the special values, everyday images and hidden meanings which are evoked by regional landscapes and places? How have these responses changed over time? Who holds them today and how do they vary across socio-economic and within professional groups, over life cycle, and among visitors, recent residents and long established inhabitants? How are landscape tastes and values acquired, transmitted and transformed? What is the social and psychological impact of physical change or threat to highly valued landscape features? What are the options for mitigating, compensating or otherwise offsetting these?

4. Institutional and Political

In landscape evaluation, there is a characteristic gap between analysis and action. The reasons why relatively little attention is paid to such research in Canada have already been noted. Much of this has to do with the policy and institutional frameworks for resource management and development planning established by the federal and provincial governments. It is also worth noting however, that a case for incorporating landscape values into environmental decision making has yet to be convincingly made, let alone demonstrated. Environmental impact assessment, for example, serves as one widely utilized and publically visible opportunity for this course of action. The process reforms presently under consideration by the Federal Government in response to the global imperative for sustainable development may be a window of entry for linking aesthetics to the proposed integration of ecology and economy.

How might landscape evaluation be institutionalized in environment and development decision making? What are the possibilities for including this approach in regional and resource planning, environmental impact assessment and related processes of project analysis? How can we build on and extend procedures which are already in place for this and similar purposes? What criteria might be adopted, for example, to environmental screening of aesthetic and visual effects? What could be the basis and approach for an in-depth visual impact assessment of a major development project not presently subject to this form of scrutiny?

5. Professional and Organizational

The final order of business on the research agenda is the procedural matter of how to pursue landscape evaluation studies in Canada. It involves, firstly, geographers interested in this specialization organizing mechanisms for information exchange and research collaboration. A special landscape research working party of the Canadian Association of Geographers may be one option. Obviously, there are many other possibilities. It may be useful, secondly, to review these as part of the broader network of professional contacts. Landscape evaluation is an area of interdisciplinary research and practice. To date in Canada, there is nothing comparable to the Landscape Research Group in the U.K. or to U.S. organizations such as the Environmental Design Research Association.

With is in mind, there are two questions left: whether a similar network needs to be put together in Canada, and if so what form might it take? It is our own conviction:

1. that the landscape values of this country are worth studying and documenting both in their own right and because of what is happening to them through ill-considered planning and development; and

2. that geographers have much to contribute directly and through a working association with other scholars and practitioners.

PLATE 25 Patterns on a sandy beach.

PHOTOGRAPHIC ESSAY

HERITAGE RETAINED: MAN AND LANDSCAPE IN BANFF NATIONAL PARK 1883-1914

Photographs of an Era from the Whyte Museum of the Canadian Rockies Banff, Alberta

with an introduction and accompanying text by
Barry Sadler, Institute of the NorthAmerican West

This collection of photographs formed part of a special exhibition to celebrate the Centennary of Canada's National Parks. The exhibition was organized by the Institute of the NorthAmerican West and was held in conjunction with the *Heritage for Tomorrow* assembly, at the Banff Centre, September 4th to 8th, 1985. It was designed "to exemplify the importance of photography as a documentary source of man's use of and response to the landscape in the early days of Banff National Park." An expanded introduction and interpretive text has been prepared from the original exhibition didactics and brochure, *The National Park Idea, the Mountain Landscape and the Visual Arts*. (The Institute of the NorthAmerican West and the Banff Centre, 1985)

INTRODUCTION

Man-environment interactions are mediated by culture and cognition. We sense and respond to the material world through a screen of knowledge and values, belief and emotion. Geographical paradigms, world views and traditional cosmologies are all ways of systematically organizing environmental and spatial experience, memory and imagination. Landscape evaluation in the broadest sense of the term, involves a certain stylized way of seeing and ordering geographical reality. The idea of landscape, as structured, expressive form, a source of beauty and meaning in its own right rather than just a setting for moral purpose or heroic acts, is a relatively recent aesthetic tradition.[1] It encompasses a range of formal themes and idealized images, which are both derived from and reflected in the course of change in cultural landscapes.

National parks are special places, set aside to preserve landscape in its natural state (which is, of course, a dynamic rather than static condition). In these reserves, the imprint of aesthetic intentions and taste tends to be clearer than in more utilitarian landscapes (where the association of form and function is more multi-faceted and hence cryptic).[2] This is still a relatively little worked area of interest, and a history of the national park movement in these terms remains to be written. A number of connections with aesthetic traditions have been traced, notably with respect to American national experience.[3] These are sketched below to provide a context for the photo-exposition of man and landscape in the early days of Banff National Park.

The very act of national park designation involves evaluating landscape and structuring space, differentiating natural areas worth formal protection from those where use and development can proceed subject to less stringent controls. A range of emotions that scholars include within the aesthetic are attached to and derived from national parks and equivalent areas.[4] Over the last one hundred years, the set of organizing values and meanings has changed: it has shifted from the scenic to the ecological; from the visual and touristic to the bio-ethical and symbolic; from unique landscape to remnant wilderness and natural area representation; and from the surface of the earth to the web of life and evolutionary processes. The course of this transition roughly parallels that from romanticism toward realism and abstractionism in the visual arts.[5] We move, in the process, from the perception of nature as the sublime manifestation of God's handiwork to a recognition of the limits to growth and the environmental impact of human activity.

Now, as then, national parks are a product of their time and place, reflecting the circumstances and attitudes of the society which utilizes them. On the North American continent, where the national park idea began and has reached its fullest institutionalization, the landscape which was the physical stimulus and symbol for the conservation movement has been itself transformed.[6] In this respect, the present American and Canadian national park systems are prophecy confirmed. They are the last islands of the biotic diversity that was once an unbroken land. Much has gone. No living man will see again the long-grass prairie stretching from horizon to horizon.[7] Each year, the stock of wild and roadless area is whittled down. Within the national parks, too, much has changed, often paradoxically so. Many landscapes, for example, have been altered subtly and, sometimes, substantially by the application of policies designed to "protect" them. Each age has a characteristic approach to balancing preservation and use, based on evolving views and values about man and his place in the natural scheme of things.

The focus of interest here is on the origins and formative years of Canada's first national park. Our memory of what we have done to such landscapes is incomplete and short.[8] Man and his artifacts have been part of Banff National Park since its establishment, and indeed since time immemorial. Stoney Indians traditionally used and occupied parts of what is now protected terrain, off limits to hunting and gathering. They consider the Cave and Basin, where the Canadian National Park system began, a sacred site.[9] One century later, it has become a historic landmark with a different set of meanings for the mainstream of society. It does not end there. As Banff enters its second century, Canadian society is very different from the one which founded national parks and is changing fast.

So the questions begin about past time and present place, about small beginnings and the widening circles of consequence, and about the sequential values associated with the nuclear site and the surrounding landscape of what is now Banff National Park. Why was this unprecedented reservation established? How, in fact, was it managed and used? What was the impact of tourist and other activities on the park landscape and their implications for the future? Banff was, and still is, Canada's premier national park (in a system which now reaches from sea to sea and encompasses some of the most spectacular landscapes and least trammelled ecosystems on the face of the Earth). The formative years, 1885-1917, were a template and testing ground of evolving policy, a mirror of land use change, and a foreshadow of future potentials and tensions.

These themes are explored here through the medium of historical photographs. As visual documents and sources of information, photographs freeze events in time, making past reality both retrievable and evocative. Early photographs of man and landscape in Banff National Park are a *heritage retained*; both for what they tell us and for what they are. With photographs, the content of the image, what it is of, is always of primary importance, (and the formal qualities of style, which are central in painting, are a secondary matter).[10] This is not, of course, to say the selection and composition of photographs is unencumbered by aesthetic values. Some of the images contained here, for example the frontispiece (p. 272) by Elliot Barnes, are classic scenes of man and landscape, powerful and economic vignettes of national park experience at the turn of the century. Old photographs also acquire their own fragile beauty, the patina of age that is a touchable link with things past.[11]

The windows which these photographs open on man and landscape in Banff National Park are organized into four thematic and chronological categories. Each one corresponds to a chapter in the evolution of the park, and to a facet of interaction. As a visual memoir, the emphasis is on images of exploration, tourism development, visitor activity, and landscape change. The text merely amplifies the dialectic of man and landscape, contrasting ideas and reality, attitudes and action. During much of the period in question, few photographs of nature were taken without including a human figure or artifacts.[12] Their presence gave visual scale and/or symbolic meaning to what was often a vast and alien space. Subsequently, the human element, heroic or otherwise, became less prominent and landscape became wilderness — but that is another story. This one begins in the United States, where the national park idea was first articulated and given visual and aesthetic form by the work of landscape artists and photographers.

BACKGROUND: FOUNDATIONS OF AN IDEA

The national park idea was forged in the firs of Yellowstone, propagated by an earlier seed from the big trees of Yo-Semite (as it was then known). It was a unique social, cultural, and political response to the American encounter with new and unusual landscapes, the visual climax of a regional geography itself dramatically different from anything else on the North American continent.[13] As the frontier ran its course, from eastern seaboard, through Appalachian Mountains, to Great Plains, a new world was discovered and created. Yet nothing prepared the society of the day for what was to come in the trans-mountain west, that "awesome space" between the Rockies and the Pacific.[14] The discovery of the natural and scenic wonders of Yellowstone, for example, was initially rejected as fiction. Once the fabled Coulter's Hell, a half century of accumulated rumours, became the authenticated fact of Yellowstone's geothermal basin, it fired the imagination of the eastern intelligensia and establishment.

By act of Congress, in 1872, Yellowstone was established as the world's first national park. An 8,990 square kilometre area was withdrawn from sale, squatting or settlement and dedicated for public use and enjoyment. This exercise in landscape democracy, at first glance, was an extraordinary departure from the frontier ethic in which economic progress was equated with private exploitation. It came at a time when Indians were still a "threat" in the Yellowstone region.[15] Only the designation of Yosemite as a state park eight years earlier served as a precedent: Yellowstone as a federal land became national real estate. In both cases, the reserve was quite different in scale and purpose from other parks of the time, whether rough common or formal garden.

Exactly why Congress passed the Yellowstone enactment, relatively smoothly and quickly it should be added, remains open to historical question.[16] The reasons are several: the first-hand enthusiasm of members of the Washburn-Langford-Doane and Hayden expeditions of 1870 and 1871; the opportunistic interest of the Northern Pacific Railroad with an eye to future enterprise; and the support of powerful and influential men in Congress. A broader set of forces were also at work, creating the socio-political climate for the national park idea. The grounds for these were the landscape itself, nature paradisical rather than primeval, complete with geological curiosities (i.e. volcanic phenomena). Yellowstone, and Yosemite before it, was both a monument and a catalyst to three themes:[17]

1. *landscape threatened* — the frontier was coming to an end and new attitudes toward nature, land and resources were emerging;
2. *landscape transcendental* — the inspirational qualities of the natural landscape were idealized in literature and art; and
3. *landscape triumphant* — there was a continuing search for the aspects of American experience which reinforced national identity and cultural distinctiveness.

A landscape artist and photographer respectively played influential roles in the establishment of Yellowstone National Park. The Hayden expedition of 1871 included the painter Thomas Moran and the photographer William Henry Jackson. Moran's massive panorama of the Yellowstone valley dramatized the quality of what was and is a spectacular vista. His canvas is widely credited with capturing the attention of Congressmen. This was, in fact, completed after the legislation, and it was Moran's watercolour sketches which were judiciously distributed to oil the wheels of passage.[18] A collection of Jackson photographs was also discreetly placed for viewing by members of the Senate. These works may have been more effective than is conventionally realized. What they furnished was a form of image truthing, a substantiation of the wonders described by explorers and portrayed by artists.[19] Only photographers could render the contemporary aesthetic into visual reality and thereby open the way to popularizing landscape wonders and a paticularized response to them (one which Paul Shepard claims owes more to intimations of the pastoral than to the appreciation of wilderness and so fits ill with the ecological realities of today).[20] For all these reasons, the national park idea, and its cultural and aesthetic rationale became formalized a decade or so before the C.P.R. line was pointed towards the east portal of the Rocky Mountains.

C. P. R. MEN

Landscape Discovered and Designated 1883-1887: The Great Railway, the Rocky Mountains and the First National Park

These are great themes of the age, part and parcel of the National Dream — tying the country together and building a nation in face of an overwhelming geography and an underwhelming history. *A Grand and Fabulous Notion*, Sid Marty calls the founding of Banff National Park and the Canadian system.[21] So it is. But the beginning was prosaic rather than heroic, especially by comparison to the history of Yellowstone. It is an early metaphor of separate identity in the face of manifest destiny; landscape discovery and designation across the 49th Paradox.

The origins of Canada's first national park are caught in the accompanying photograph of the transcontinental railway being driven toward the Continental Divide. A new and spectacular landscape is opened, overflowing with resource wealth to be harnessed for the "purposes of the Dominion".[22] One of the ways this would be done is pre-shadowed in the weary and quizzical faces of the railway workers, here pausing rather than posing at their labours. In November 1883, three of their colleagues, Frank McCabe and William and Thomas McCardell, prospecting in the vicinity of Banff Station (then Siding 29) found or rediscovered the thermal pool now known as the Cave and Basin and subsequently located the Upper and Middle Hot Springs on the slopes of Sulphur Mountain.

During the next two years, the two were joined by "a powerful aggregation of manpower" in laying claim to the Springs.[23] Several rudimentary improvements were made to the Cave and Basin by the McCardells, McCabe and others. By the summer of 1885, more substantial entrepreneurs had begun taking a speculative interest in the area. It was this situation which apparently precipitated the decision to establish the Banff Hot Springs Reservation on November 25, 1885. Order-in-Council 2197, withdrew from sale, settlement or squatting a 26 square kilometre area encompassing "several hot springs which promise to be of great sanitary advantage to the public".

Hardly a ringing declaration of high purpose but the intent was clear enough. With the encouragement of the C.P.R., in the shape of William Van Horne, the first "Lord of the Line", the Government proposed to develop a spa resort at Banff bringing customers to the railway and "prestige to the whole country".[24] A survey of the Reservation and plans for a townsite, modelled after Hot Springs, Arkansas, were prepared in 1886 by the first superintendent, George A. Stewart. His initial report drew attention to the beauty of the surrounding landscape, and he was subsequently asked to survey a wider area and included "all reasonable points of interest".

The following year, the Hot Springs Reservation and the adjacent townsite of Banff were incorporated within Rocky Mountains Park, a 673 square kilometre rectangle which served as testimony to Stewart's early training as an engineering surveyor. In drafting the establishing Act, the legislators borrowed freely from the language of the Yellowstone Park Act of fifteen years earlier. However, there were also subtle yet decisive differences between American and Canadian approaches to the institutionalization of the national park idea; which, in part, reflect their different stages of frontier evolution.

From the outset, it was apparent that the primary purpose of Rocky Mountains Park (like Yellowstone) was the encouragement of tourism. The protection of landscape was one of the means to that end. Scenic values as well as "sanitary advantage", the reputed therapeutic powers of the Hot Springs, were envisaged as the main assets of the newly established park. As an unprecedented form of landscape designation in Canada, Rocky Mountains Park and its immediate successors in the region should be seen as an extension to rather than a departure from the "doctrine of usefulness", the vigorous pursuit of resource exploitation in the national interest.[25] This policy thrust was quite evident in the House of Commons debate of the park bill. Most of the opposition to the proposal was on the grounds of government involvement and expenditure and not against the idea *per se*. Only one or two members noted the potential contradictions between the primary purpose of the reserve and the provisions for continuing other exploitive activities (unlike Yellowstone). Such a view, logical enough today, must be placed in the context of contemporary perceptions and assumptions, of the Great Lone Land and an inexhaustible supply of resource wealth. In the years which immediately followed, images and words were backed by deeds and practices.

Landscape Improved and Tamed, 1887-1910: Park Making, Resort Development and Tourism Activity

The quarter century following the establishment of Rocky Mountains Park was one of considerable resort development at Banff and its environs. A range of 'improvements' to the landscape — 'park making' — were undertaken. It was part of the grand strategy of Van Horne's C.P.R. to "capitalize the scenery".[26] The provision of a range of facilities, rather than the maintenance of wilderness was required for this purpose; only with such development would the landscape become a park.

Nor was Rocky Mountains Park envisaged as being 'for all people'. The emphasis was on developing a fashionable watering place and alpine resort for a well-heeled, international clientele. It was manifestly not meant to encourage "a doubtful class of people", which would certainly include "the fashionable and crowded population situated between Callandar and Rat Portage."[27] The Rockies, the singing mountains and the unbroken silence

were proclaimed as "the Mountain Playground of the World". Early advertising had an exaggerated flair, backed by a shrewd investment in facilities and people, so that over time the precocious image became contemporary reality.

A cornerstone of the era of park making was the C.P.R.'s construction of a grand hotel at a site overlooking the confluence of the Bow and Spring Rivers. The Banff Springs Hotel, was completed in the Spring of 1888, and became the touristic motif of the new park.[28] It was (and still is) geographically and symbolically set apart from the townsite of Banff, then a rude collection of a dozen or so buildings. Both unique and memorable, "the Springs" was *The Hotel*, advertised by the C.P.R. as the "Finest on the North American Continent", a source of plush comforts and cosmopolitan pleasures for its wealthy guests and a romantic and picturesque landmark for other early tourists. In architectural style, the original hotel, shown with later additions in the accompanying photograph, tended toward a French Chateau exterior and a Scottish Baronial interior.

Soon a growing number of other accommodations were competing with each other, often vocally, for the business of early tourists stepping down from the train at Banff Station. A line of horse-drawn tally-hos conveyed visitors to their hotels and places of interest in and around the townsite. The initial development activities of the Park Administration had focussed on the construction of a network of coach or tote roads and rustic facilities at the Cave and Basin and Upper Hot Springs. By the turn of the century, government activities extended to the provision of recreational "attractions" (including a boat house, animal paddock and a zoo and aviary) and soon after to the establishment of municipal-type services.[29] During the perod 1887 to 1910, a range of landscape improvements were centred around the removal of fallen timber, and other "unsightly objects" and on planting additional trees, shrubs and plants.

Private enterprise also played a role in shaping the face of early tourism, adding local colour and human drama to the formative history of Banff and Rocky Mountains Park. The large, and not always harmonious, association of the Brewster family and the National Park, for example, dates from the time of its establishment.[30] A dynastic family of pioneering outfitters and modern transportation czars, the Brewsters and their business partners began to exert a growing influence on patterns of park land use and tenure in the new century. With the acquisition of the C.P.R.'s livery concession in 1904, the Brewster stables became the hub of a daily round of tourist activity. The impressive new building and the complement of carriages, horses and guides, illustrated on page 283, speaks of both the trappings of frontier tourism and corporate prosperity in the making.

During the first decade of the twentieth century, the number of people visiting Rocky Mountains Park increased roughly ninefold (from approximately 6,500 to 56,000). The scope of their activities also diversified in

BREWSTER TRANSFER LIVERY

range, distribution and seasonality. Three photographs are juxtaposed here to convey the sense of touristic time and place: bathers and bystanders at the Cave and Basin (above); ladies dressed in formal attire for horse riding along Lake Louise trails (p. 285); and the more robust pleasures of ice yachting on Lake Minnewanka (p. 286). Landscape tamed by activity and association; resort in the wilderness; tourism as frontier enactment — these are the enduring images of early Banff and environs.

Landscape Exploited and Explored, 1887-1911: Nature, Wildlife and Other Resource Uses

Beyond the townsite, the surrounding landscape was certainly not "wilderness throughout" even at the time of the establishment of Rocky Mountains Park.[31] Forest fires had markedly increased since the coming of the railway and large tracts of the Bow Valley were already burnt over by

1887; mining and lumbering concerns (see above and p. 288) also predated park designation; and, as a result, game and fish had become "comparatively scarce". Landscape change continued throughout the formative years of the Park. The 1887 Act contained clauses providing for resource protection, but the appropriations and manpower were insufficient and the pervading attitude toward nature, wildlife and land use were both ambivalent and ambiguous.

Early policy toward the control of forest fires, for example, was necessarily passive, confined to the cutting of a few fire breaks which did nothing to prevent major burns in the period between 1887 and 1911. The impact on natural processes of vegetation succession were considerable, with large increases in fire-following lodgepole pine and the expansion of grasslands.[32] Park wildlife was also affected by these events and the ravages of hunting. In 1890, the Government had banned big game hunting in Rocky Mountains Park, except for large predators (avian as well as terrestrial) and certain animals regarded as 'nuisances'. Banff continued to develop as an outfitting centre for hunting trips beyond park boundaries, although this became more problematic after the 1902 expansion.

The massive increase which took place in the size of Rocky Mountains Park in that year was apparently designed to expand tourism opportunities and revenues and to permit more effective wildlife management. In fact, the new configuration of the park was little more rational than the old one and a great deal more difficult to police and control. During the years 1887 to 1911, poaching was undoubtedly widespread; on occasion, the Brewster family and other Banff outfitters "encountered difficulties with the Game Laws".[33] For much of this period, the captured trophy and the performing bear (pp. 289, 290) were more characteristic images of man-wildlife relationship in the Rocky Mountains Park than predator-prey interactions and their place in the ecological system.

Other natural resources were also exploited locally in Rocky Mountains Park between 1887 and 1911.[34] Coal mining at Anthracite and timber cutting in the Spray Valley were included within the initial park boundary. The 1902 extension encompassed further 'non-conforming land uses', including the mining settlement of Canmore. By contrast, the opening of a C.P.R. mine at Bankhead (p. 287) in 1904 was a park intrusion rather than inheritance. It was, moreover, a highly visible activity, located on the tourist drive from Banff to Lake Minnewanka. The Eau

Claire Lumber Company (p. 288) continued cutting much undisturbed timber in the general vicinity. Both mining and timber activities created industrial scenes which were more in keeping with a frontier rather than a national park ethic and aesthetic. At the time, however, tourist response may have been less negative. Park Superintendent Douglas even considered Bankhead and "its teeming industrial life" as one of "the attractions of the neighbourhood". His boosterism eventually came partly true: Bankhead is now a ghost town reclaimed by nature, a self-guiding walkway leads the visitor through the silent ruins.

Not all tourists of the period were confined to the tote roads and riding trails of Banff and Lake Louise. The Brewsters and other outfitters took small but increasing numbers of sportsmen and other visitors into the back country, either following game trails or blazing their own. A larger than life cast of characters, like Tom Wilson, Jimmy Simpson and Bill Peyto, acted as guides and packers to both city dudes and more serious explorers (p. 292).[35] By the turn of the century, Banff had become an important centre for mountain climbing, the unclimbed and often unnamed peaks of the Rockies, became the target for ascent by a small group of celebrated alpinists.[36] Swiss guides, such as Edward Feuz, Rudolf Aemmer and Konrad Cain, were imported by the C.P.R. to help climbers and hikers reach their destination safely. In the photograph (p. 293), they gaze diffidently out at a landscape explored under their tutelage, and perhaps look to other areas still "awaiting the first footfall".[37]

Landscape Conserved and Rationalized, 1911-onward: The Conservation Movement and Portents of Change

In 1911, the boundaries of Rocky Mountains Park were adjusted to coincide with the watershed of the Bow River. This change, while it involved a considerable reduction in park size, signalled the start of the conservation era. A number of progressive measures for landscape and wildlife protection were instituted in years immediately preceeding the 1911 boundary revision. It was, however, the rationalization of the boundary with a natural unity, followed the next year by the appointment of J.B. Harkin as the new Commissioner of Dominion Parks, that marks the opening of a new administrative chapter in Canadian national park history.[38]

Under the influence of the Conservation movement, there was a growing recognition of the interdependencies of resource uses, and their implications for the park landscape. The policy pathway begun at this time led eventually to the benchmark National Park Act of 1930, which still remains the philosophical mainstay of resource use and management. *En route*,

resource extraction and other non-conforming activities were phased out of Rocky Mountains (later Banff) National Park. The dual mandate of tourist use and enjoyment and landscape preservation "unimpaired for future generations" contained its own inherent tension, though this was not widely perceived at the time.[39]

The harbingers of coming change in the final two photographs seem innocent enough: several cars parked in Banff townsite just before World War One (p. 295) and a landscape photographer pursuing his art (p. 296). Yet their combined impact on park use and development was quite revolutionary. Rocky Mountains Park was established in Van Horne's memorable phrase, to import the visitor, "since we cannot export the scenery."[40] The private automobile was to change the basis of "importing visitors"; the photographic image became a *de facto* "exporter of scenery" to a wide and eager audience.

The first car to make the journey into Rocky Mountains Park in 1904 did so without the benefit of roads or official sanction. An Order-in-Council prohibiting automobile access was passed the following year, but partially amended in 1910. Thereafter capitulation was swift: by 1915, automobiles were permitted on all park roads and the horse-drawn tally-ho was rapidly replaced by motor bus service (operated, of course, by the Brewster family). During the inter-war years and into the modern era, the personal mobility conferred by growing automobile ownership made the park increasingly accessible to visitors. It was the car which was to translate the altruistic idea of national parks for all people into reality. Landscape democracy, however, had its price: the demands of the automobile visitor was to remake the park landscape. More than any other factor, the car was the catalyst of the tension between use and preservation as the trickle of visitors eventually became a flood.[41]

Another catalyst, less easily quantified, was the "export of scenery", especially via the postcards and photographs of Byron Harmon. He was to introduce Banff National Park and the Rocky Mountains to the world. Much of his work was undertaken during major expeditions. Between 1906 and 1913, for example, he was official photographer of the Alpine Club of Canada and recorded all of the early camps and many famous ascents. This involved packing the heavy and awkward photographic paraphenalia of the 1900's, including after 1910 a movie camera.[42] In the photograph, Harmon is himself caught in the act of 'shooting' the mountain landscape he loved and idealized. The endpiece (p. 302) is one of the more dramatic of several thousand images he captured in the Rockies during his lifetime, much of it spent waiting for the short, often fleeting moment when the play of light and shadow was right.

A CONCLUDING NOTE ON LANDSCAPE PHOTOGRAPHY AND NATIONAL PARKS

Landscape photography and the National Park idea have evolved together, one has reinforced the other. This relationship is relatively well documented in the United States with respect to both the establishment and early history of National Parks and to their more recent evolution.[43] It is much less known in Canada. A number of publications have appeared recently on the life and work of photographers of the Rockies, notably Elliot Barnes and Byron Harmon.[44] The impact on the national park, as idea and institution, has not been explicitly scrutinized (or *vice versa*). It is a story worth telling; the images and text hint at themes to be uncovered.

REFERENCES

1. SZARKOWSKI, J., *American Landscapes.* New York: The Museum of Modern Art, 1981, p. 5.

2. SADLER, B. and CARLSON, A.A., "Environmental Aesthetics in Interdisciplinary Perspective", in SADLER, B. and CARLSON, A.A., (eds.) *Environmental Aesthetics: Essays in Interpretation.* Western Geographical Series Vol. 20, University of Victoria, 1982, p. 8.

3. See, for example, HUTH, H. *Nature and the American: Three Centuries of Changing Attitudes.* Berkeley: University of Calf. Press, 1957, especially at pp. 148-164; and RUNTE, A., *National Parks: The American Experience.* Lincoln: University of Nebraska Press, 1979.

4. The nature of aesthetics, and its relationship to nature, environment and landscape is the subject of much discussion by philosophers and others. See, for example, HEPBURN, R.W. "Aesthetic Appreciation of Nature", in OSBORNE, H., (ed.) *Aesthetics in the Modern World.* London: Thomas and Hudson, 1968, pp. 49-66. A trenchant critique of the philosophical basis of landscape evaluation research is in CARLSON, A.A., "On the possibility of quantifiying scenic beauty", *Landscape Planning*, 4, 1977, pp. 131-172.

5. MCSHINE, K. (ed.), *The Natural Paradise Painting in America 1800-1950.* New York: The Museum of Modern Art, 1976.

6. TREFETHEN, J.B., *The American Landcape: 1776-1976, Two Centuries of Change.* Washington D.C.: The Wildlife Management Institute, 1976.

7. LEOPOLD, A., *A Sand Country Almanac and Sketches Here and There.* New York: Oxford Univesity Press, 1987, p. 189.

8. NELSON J.G., *The Last Refuge.* Montreal: Harvest House, 1973, p. 1.

9. KAQUITTS, F., "The Role of Indians in the National Park System", in SCACE, R.C. and NELSON, J.G. (eds), *Heritage for Tomorrow*, Vol. 5. Ottawa: Supply and Services Canada, 1987, p. 226.

10. SONTAG, S., *On Photography*. New York: Parrar, Straus and Giroux, 1973, p. 93.

11. CAVELL, E., *Sometimes a Great Nation, A Photo Album of Canada 1850-1925*. Banff: Altitude Publishing Ltd., 1984, p. 7.

12. *Ibid.*

13. See SHEPARD, P., *Man in the Landscape*. New York: Ballantine Books, 1967, pp. 235-271.

14. This and other themes in past and contemporary human interactions with the intermountain landscape are traced in HART, E.R. (ed.), *That Awesome Space*. Salt Lake City: Westwater Press, 1981.

15. Chief Joseph led the Nez Pearce through the Park on their long retreat toward the Canadian border and his famous recognition of inevitable destiny: "I will fight no more, forever". The self-congratulatory theme which often creeps into discussions of the cultural altruism behind national park establishment on the American frontier ignores the dark shadow cast by the previous sad history of native dispossession.

16. See BARTLETT, R.A., *Nature's Yellowstone*. Albuquerque: University of New Mexico Press, 1974, pp. 194-210.

17. *Ibid.*; SHEPARD *op cit.* and RUNTE, *op cit.*

18. BARTLETT, *op cit.*, p. 208.

19. OSTROFF, E. *Western Views and Eastern Visions*. Washington, D.C.: U.S. Government Printing Office, 1981, p. 14.

20. SHEPARD, *op cit.*, pp. 247-265.

21. MARTY, S., *A Grand and Fabulous Notion*. Toronto: NC Press, 1984.

22. The phrasing is that of Prime Minister John A. MacDonald. It conveys the political intent of the period, and the strong link that was made between park establishment and the development of the Canadian nation (then still young). For a discussion of this relationship, see: BROWN, R.C., "The Doctrine of Usefulness: Natural Resource and National Park Policy in Canada, 1887-1914", in

NELSON, J.G. and SCACE, R.C., (eds.), *The Canadian National Parks: Today and Tomorrow*. Studies in Land Use History and Landscape Change, National Park Series No. 3, The University of Calgary, 1968, pp. 94-110.

23. Marty, *op. cit.*, pp. 33-44, covers this story in some detail and with imaginative flair absent from more conventional histories. See also LOTHIAN, W.F., *A History of Canada's National Park*, Vol. I. Ottawa: Parks Canada, 1976, pp. 18-22.

24. John A. MacDonald again. For a discussion of the spa proposal and its implementation, see LOTHIAN, *op. cit.*, Vol. 3, 1979, pp. 12-25; and SCACE, R.C., *Banff: A Cultural Historical Study of Land Use and Management in a National Park Community to 1945*, Studies in Land Use History and Landscape Change, National Park Series No. 2, University of Calgary, 1968, pp. 19-65.

25. BROWN, *op. cit.*

26. ROBINSON, B., *Banff Springs: The Story of a Hotel*. Banff: Summerthought, 1973, pp. 3-5.

27. "Doubtful class of people" was another MacDonaldism; it can be roughly translated as "ordinary people". The accompanying general view of Western Canadians was contained in the *Toronto Globe*, July 13, 1886. (It reflects, of course, a media trait that many westerners contend continues to this day).

28. See ROBINSON, *op. cit.*

29. Full details of resort development and other aspects of land use and settlement history in this period are contained in SCACE, *op. cit.*; and LOTHIAN, *op. cit.* (Vol. 3). See also LUXTON, E.G., *Banff, Canada's First National Park*. Banff: Summerthought, 1975, pp. 90-108.

30. HART, E.J., *The Brewster Story*. Banff: Brewster Transport Company Ltd., 1981.

31. BYRNE, A.R., *Man and Landscape Change in the Banff National Park Area Before 1911*. Studies in Land Use History and Land-

scape Change, National Park Series No. 1, University of Calgary, 1968.

32. *Ibid.*

33. HART, *op. cit.*

34. See BYRNE, *op. cit.*, pp. 119-124; LUXTON, *op. cit.*, pp. 85-89.

35. FRASER, E., *The Canadian Rockies*. Edmonton: Hurtig Ltd., 1969, pp. 119-230.

36. DOWLING, P., *The Mountaineers*. Edmonton: Hurtig Ltd., 1979.

37. The phrase is Norman Collie's, one of the Grand Old Men of Rocky Mountaineering. See TAYLOR, W.G., *The Snows of Yesteryear*. Toronto: Holt, Rinehart and Winston, 1973.

38. An administrative and policy history of Rocky Mountains and other National Parks is in LOTHIAN, *op. cit.*, Vol. 2, 1976, especially pp. 10-17.

39. SADLER, B., "Bordering on the Magnificent", *National Parks*, 59, 7/8, 1985, p. 27.

40. Quoted in Fraser, *op. cit.*, p. 118.

41. SADLER, B., "The Machine in the Wilderness: Automobiles, Roads, and the National Parks", *The Albertan Geographer*, 10, 1974, pp. 66-70.

42. HARMON, C. (ed.), *Great Days in the Rockies: The Photographs of Byran Harmon, 1906-1934*. Toronto: Oxford University Press, 1978.

43. CAHN, R., and KETCHUM, R.G., *American Photographers and the National Parks*. New York: The Viking Press, 1981.

44. HARMON, *op. cit.*; and CAVELL, E., *A Delicate Wilderness, The Photography of Elliot Barnes, 1905-1913*. Banff: Altitude Publishing, 1980.

THE CONTRIBUTORS

PHILIP DEARDEN is an Associate Professor of Geography at the University of Victoria in British Columbia. His main research interests focus on man-land interactions particularly related to conservation, resource-based tourism and landscape aesthetics. He has published extensively in the latter field with most emphasis on methodological questions.

JOHN E. FITZGIBBON received his B.A. from McMaster University, M.Sc. from the University College of Wales (Aber.) and Ph.D. from McGill University. He has taught geography at the University of Saskatchewan and University of Guelph, and is currently jointly appointed between the School of Rural Planning and Landscape Architecture at Guelph. His current research is in the areas of landscape assessment, environmental planning and water resources planning.

EDWARD M. W. GIBSON is the Director of Simon Fraser Gallery; and Associate Professor, Department of Geography; and an Associate, Center for the Arts Simon Fraser University, Burnaby, B.C. His publications and video productions are in Humanistic and Aesthetic Geography and he has served as an appointed member of heritage and art advisory boards at the municipal and provincial levels in British Columbia.

MILFORD B. GREEN is an Associate Professor of Geography at the University of Western Ontario. He has published over 50 articles on a diverse set of topics ranging from the Soviet Union, the South Pacific, the Northwest Territories, mergers and acquisitions, joint ventures, venture capital, and rail transport rates. A long-term interest has been in corporate geography, a subfield of economic and industrial geography. His major focus has been interfirm relationships, particularly in Canada and the United States.

LOUIS HAMILL earned his doctorate in economic geography at the University of Washington in 1963. He taught at the University of Calgary from 1963 to his retirement in 1988. His teaching and research centred on environmental analysis, environmental and resource management, environmental problems, urban and non-urban recreation, and tourism. Techniques of recreation resource, scenery and recreation demand analysis were among his special teaching and research interests.

ROBERT M. ITAMI is an Assistant Professor of Landscape Architecture in the School of Renewable Natural Resources at the University of Arizona in Tucson, Arizona. He received his B.L.A. from the University of Idaho, and his M.L.A. from the University of Melbourne, Australia. Itami has been involved in visual resource research, teaching and consulting in the United States, Australia and Canada for the past 10 years. His current research focusses on improving the performance and reliability of visual assessment procedures using geographic information systems and other computer-assisted analysis techniques.

JOHN MARSH obtained his Ph.D. from the University of Calgary in 1972 studying the historical geography of Glacier National Park. He is now Professor of Geography at Trent University. He has continued his research on the historical perceptions of park landscapes and their portrayal in tourism advertising. Recent work has focussed on the portrayal of Canada in Japanese tourism advertising and the evolution and perception of St. Andrews, New Brunswick as a resort.

MICHAEL R. MOSS is Professor and Chairman of the Geography Department at the University of Guelph. His main teaching and research interests lie in the field of the biophysical analysis of landscapes and the application of biophysical process formation to procedures of land management and environmental impact assessment.

WILLIAM G. NICKLING is a Professor in the Geography Department at the University of Guelph. His main research interests focus on the measurement and assessment of soil erosion by wind with particular reference to agricultural soils. This work also encompasses the development of control strategies and land management practices to decrease wind erosion on susceptible soils.

EDMUND C. PENNING-ROWSELL is currently Professor of Geography and Planning, and Dean of Social Science, at Middlesex Polytechnic, London, England. He graduated from University College London, with a B.Sc. (Geography) in 1967 and in 1970 with a Ph.D. (Geomorphology). He also later took an M.A. in Social and Industrial History, in 1986, from Middlesex Polytechnic.

His main research interests are in the field of water resources and landscape perception. He is the author of *The Benefits of Flood Alleviation* (1977), *Water Planning in Britain* (1980), *Weather and Water* (1985), *Landscape Meanings and Values* (Ed.) (1986), *Floods and Drainage* (1987), and *Risk Communication and Response* (Ed.) (1989), plus some related papers on floods, water, landscape perceptions, and geomorphology.

He is Head of the Middlesex Polytechnic Flood Hazard Research Centre and currently adviser to Thames Water, the O.E.C.D., the United Nations Disaster Relief Organization, the British Department of the Environment, the Welsh Office and a small number of more interesting amenity societies.

JOHN W. POMEROY studied landscape aesthetics from 1981 to 1984 in the Geography Department at the University of Saskatchewan, Saskatoon, as part of a programme to specify the environmental and recreational resources of the South Saskatchewan River. In 1988 he received a doctorate at the Division of Hydrology, University of Saskatchewan, based on an investigation of blowing snow on the Canadian Prairies. He is a hydrologist with the Rocky Mountain Forest and Range Experiment Station of the U.S. Forest Service in Laramie, Wyoming, and is presently on leave as a NATO Science Fellow with the School of Environmental Sciences, University of East Anglia, Norwich, U.K. His current studies involve blowing snow and its environmental effects, and he has a keen interest in landscape and its preservation.

J. DOUGLAS PORTEOUS became Professor of Geography at the University of Victoria via Oxford, Hull, Harvard, and the Massachusetts Institute of Technology. He currently has seven books in print, ranging from *Environment and Behaviour: planning and everyday urban life* (1977), through *The Modernization of Easter Island* (1981) to *Degrees of Freedom* (1988), a book of verse, and *Planned to Death*, a combination of history, social anthropology, regional planning critique, and autobiography. A further book, *Otherscapes: sensuous worlds and landscapes of the mind*, will appear in 1990. Dr. Porteous' research interests are eclectic, and he has written on such diverse themes as environmental psychology, urban design, urban and regional planning, geopolitics, the poetry of landscape, transportation history, literary landscapes, the philosophy of geography, and environmental aesthetics. The common bond is a radical-humanist concern for the quality of human-environment relationships.

BARRY SADLER, a former member of the Department of Geography, University of Victoria, is Director of the Institute of the NorthAmerican West, Victoria. His main areas of interest are in environmental assessment and resource management with a particular orientation to the institutional and behavioural aspects of decision making. For the past five years, Sadler has been retained by the Federal Environmental Assessment and Review Office as an advisor on policy and science and a member of the Canadian Environmental Assessment Research Council. He is also a Consulting Associate of the School of Management, The Banff Centre, a member of the IUCN Commission on Sustainable Development, and currently acts as Director of Special Projects for Globe 90.

COLIN J. B. WOOD is Associate Professor and Chair, Department of Geography, University of Victoria and former editor of *Park News*. He has published research on various aspects of resources management and cultural geography and is currently working on a project in South East Asia. He has a keen interest in landscape and landscape art.